ĒMA THURAIRAJAH

Being. Belonging. Becoming.

First published by Praxis Press 2024

Copyright © 2024 by ēma thurairajah

All rights reserved. No part of this publication may be reproduced, stored or transmitted in any form or by any means, electronic, mechanical, photocopying, recording, scanning, or otherwise without written permission from the publisher. It is illegal to copy this book, post it to a website, or distribute it by any other means without permission.

First edition

ISBN: 978-1-0689092-3-8

Contents

Preface   vii

I Being

1. Deconstruction   3
2. Where We Begin   6
3. Being Us   12
4. Our Body, Our Self   23
5. Made by Memory   35
6. Thinking Beings   44
7. Bugs for Brains   53
8. Navigating Problem Space   57
9. Feeling Like Some Body   61
10. Why Behave   65
11. Who and Whom   70
12. Where We End   81

II Belonging

13. The Centre of the Universe   89
14. The Big Picture   93
15. When are We   98
16. Entangled Forever   104

17	All That We Are	109
18	Being in the World	116
19	Being Social	126
20	We Belong to Us	134
21	On Our Own	140
22	Touched by Magic	147
23	(Be)longing to Work	151
24	We Are Here	161

III   Becoming

25	Becoming Who We Are	169
26	Letting Go	174
27	Accepting What Is	181
28	Looking at What We See	186
29	The Price of Everything	193
30	Picking the Right Horse	198
31	The Limits of Reason	203
32	Coming to an Understanding	209
33	Relating	215
34	What We Must Do	221
35	Making Meaning	226
36	A Life That's Good	235

今は今

"In many ways writing is the act of saying I, of imposing oneself upon other people, of saying *listen to me, see it my way, change your mind.* It's an aggressive, even a hostile act. You can disguise its aggressiveness all you want with veils of subordinate clauses and qualifiers and tentative subjunctives, with ellipses and evasions— with the whole manner of intimating rather than claiming, of alluding rather than stating— but there's no getting around the fact that setting words on paper is the tactic of a secret bully, an invasion, an imposition of the writer's sensibility on the reader's most private space"

— Joan Didion

Preface

What makes you, you?
You are clearly an individual, or so you say. You stand apart from me. You are you, and I am me.
But what makes this so?
Perhaps it is your identity as defined by things like name, ethnicity, citizenship, political affiliation, or religion. Or possibly it is your genes. Perhaps it is the un-shareable, subjective experience of being you.
Maybe it is all those things.
Regardless, most of us walk around with a palpable sense of self— a set of characteristics, emotions, perceptions that make us feel like we are distinct from everything around us, including our natural and social environments.
But how accurate is this feeling? How reliable? And importantly, how real?
Well, the answer to "who are you?" is not as simple as it may seem at first blush.
Yet few of us spend much time thinking about questions like these, let along digging deeper. We simply assume an identity— based on where we find ourselves, what work we do, whom we associate with. That identity gives us a sense of security to do things in the world. And when it is useful, to live life on auto-pilot— obviating the need to dive below the surface of everyday existence. Until, that is, something blows us off course and we become unmoored— unknown, even to ourselves.
The question of who we are may be as old as human consciousness

itself, although it is doubtful that a sudden, momentous spark of self-awareness triggered its asking. Perhaps we began to wonder about ourselves gradually, in a process of awakening that spanned millennia. We may have started with vague wonderment, followed by abstract notions and once we had language, moved on to a more formal formulation of such questions. We may never know for sure; we can only infer.

And so it is with other living things. Self-awareness tests often use mirrors to detect consciousness in animals. While the merits of this approach are debatable, it has thus far identified everything from birds to fish as being self-aware, in addition to dogs and monkeys, which many of us would readily acknowledge as being conscious.

Nonetheless, do animals question their identities, their subjective experience, as humans do? Or do they ponder their existence in a manner that is unique to their kind? Are self-awareness and self-inquiry even possible without language?

Whether or not they engage in self-reflection, it is now generally accepted that most, if not all, animals are conscious. At the most elementary level, if we define conscious experience as the ability to sense the natural environment and respond to stimuli from it, then it is evident that every living thing would qualify.

But there is accumulating evidence that animal consciousness goes well beyond that most basic level. Experiments have shown that animals learn, remember and act based on their environmental awareness.

However, regardless of whether it is a human being or an animal, can a thing ever truly understand itself—no matter how aware? Can brains studying themselves ever be objective? To paraphrase the physicist Max Planck, we may not ultimately be able to solve the mystery of ourselves because we are an inextricable part of that mystery.

All the same, there is value in going as far as we can, in the hope that the inquiry itself will be revelatory, with or without a satisfactory conclusion. Whatever limitations exist to self-knowledge, we can't help trying. Beginning to understand what we value; why we behave as we do; have the emotions we do, can only help us choose a life that is more fulfilling, less despairing and is more likely to improve our relationships with others.

We do still need to maintain a healthy skepticism about such efforts and findings. To Planck's point, it is important to remember that we may ultimately not have access to the unconscious (and possibly more accurate) reasons for our behaviour; and the explanations we come up with may be flawed even if they feel true.

Indeed, it seems foolhardy to pursue an all-encompassing philosophy of life in the absence of absolute and complete self-awareness. Thus, we will settle on articulating a philosophy of living—not of happiness, success, or fulfillment—but merely an approach to everyday thought and action.

At this point, it is important to acknowledge that these lines of inquiry are well worn, the subjects of philosophical and scientific study for millennia. Thus, this book is an attempt to integrate different approaches to studying the self and present a summary that is multidisciplinary and accessible to the general reader. Be forewarned, however, that getting through the book is likely to require a good measure of motivation, curiosity and patience to navigate the weft and weave of what may at times seem like dense material.

Finally, as much as I think it is important to understand ourselves better to act more purposefully in the world, we should keep in mind that in the final reckoning the answer to the question of "who am I?" may not be as important as we would like to think it is. If, after all, we are only cosmic dust accrued on so much bone, the stuff of stars given substance by the illusion of flesh; a speck of dust amongst countless billions of other specks; a moment in an infinite number of moments.

A note on the evidence

Why should you believe any of this? Indeed, why believe anything at all? Why *do* we believe or not believe? This goes to the heart of epistemology. But we do not have to get too philosophical to consider what to believe and when to believe it.

Ultimately, belief— and we are not talking about the religious kind here— is a precondition for existence. It is impossible to be and act in the world without believing something about yourself and your relationship with, and influence on, your surroundings.

So, we believe because we must. We believe to live. Unbeknownst to ourselves, we become good Bayesians— we rely on priors, on our experience of the past to inform the present and future. And, ideally, when those assumptions— when those priors—are wrong, we update our beliefs. But this does not always happen— perhaps it happens rarely, because we become invested in a particular interpretation of the past, a specific story about ourselves and the world. Yet, it happens enough for us to live another day.

This is where the scientific method is supposed to help. It does so by continually subjecting our hypothesis to testing. By remaining open to being proven wrong, we try to address our psychological failings— our need to believe without evidence.

And yet, we can only be so rational. We are not always aware of our own irrationality (even scientists, both social and yes—even physical scientists— are subject to this bias), which leads us to believe things because we want to believe them; because we want them to be so, because we want the world to conform to our mental model of it.

Thus, you should take what's presented here, especially what falls into the social science category (e.g., psychology), with a grain of salt. While some results have remained consistent through multiple experiments over many years, others are more recent and may fail replication tests. And both types may still have statistical errors and undetected biases. Further, it is important to remember that results from a controlled lab setting often can't be replicated in more complex settings, like the real world. This is especially so because many psychological experiments were conducted with so-called WEIRD (Western, educated, industrialized, rich, democratic) subjects (who were also typically college-aged, white and American).

As much as some social scientists apply the scientific method and subject their results to so-called p-value analysis (the likelihood of a result being due to chance rather than due to the hypothesized cause), the standard of proof is several orders of magnitude lower than it is for the physical sciences— especially Physics. And even something as rigorous as Physics is susceptible to errors that may be caused by noise in the data, biases, or methodological issues to do with data collection or statistical analysis.

Similarly, findings in other fields, including biology and within it, medicine, are subject to the correlation-causation fallacy, wherein two (or more) factors are assumed to be related because they are correlated in some fashion (one goes up or down at a specific rate relative to the other). This fallacy is common in small samples where random events may be interpreted as being meaningful— and the effect disappears when the experiment is done with much larger populations.

So, when we talk about human behaviour and the motivation for it; when we discuss the reasons why people are the way they are—why they behave as they do—we cannot do so with as much certainty as we have in predicting that a certain combination of chemicals will have a known set of properties or that applying a certain voltage with make a bulb light up. That is, behavioural theories are not laws of physics.

Organization

My aim in this book is to help you seek self-knowledge, not in a religious or even spiritual sense but in ways that are supported by the best evidence currently available. And we will seek to go a little beyond, to explore areas that haven't yet yielded sufficiently to scientific scrutiny, to see what philosophy has to offer in terms of useful questions and new ways of thinking about ourselves and the reality of our existence.

Our particular exploration will start from the most basic building blocks of self, namely genetics, and before branching into other areas, such as chemistry, biology, and therein to physiology and neuroscience. Then, we will look at the definitions of the self from psychological and sociological perspectives, and eventually a philosophical one, specifically when we discuss consciousness, which is the ultimate basis of self-conception.

Finally, we will discuss the end of the self (death) and consider its implications for life. It may seem like an unusual and morbid way to conclude an exploration of self-discovery, but there is more to it than that. By being among the few things that will happen to all of us with (at least for now) 100% probability, death is a shared fate that has tremendous implications for how we live.

By some accounts, our inability to accept our own mortality is what causes much of our suffering and confusion. So, could acknowledging and coming to terms with the end be an antidote; perhaps even the key to a life lived well. Could it yield a deeper understanding of the nature of existence itself? Who knows? All we can do is ask the questions, explore the ideas, and see where they lead us.

In Part II, after exploring the nature of our being, we will delve into the nature of belonging—not just in the context of family, friends, and society, but also in terms of how we relate to the planet and the universe. This necessitates a brief detour through our historic origins and into the nature of reality itself—the very thing of which we are a part.

From there, in Part III, we will consider what it means to become something other than what we already are, including questions of how we can or should be in the world, what we can use to guide our actions, and how we may be able to find or create meaning. To be clear, this section—as opposed to the preceding ones—is less a researched pontification of facts and figures and things verified, and more an exploration of ideas.

I

Being

1

Deconstruction

We walk around believing that we know something about ourselves—our fears and desires; our talents and failings; our likes and dislikes.

Even if we aren't always clear about the origins of these qualities or preferences, we believe that there is a core self that is true. We think that we are a certain kind of person, that we have a personality and that some things are in our nature or true to it, and other things are not. We are this, not that, we say— to ourselves and others.

Nonetheless, we are not always sure how we know these things, only that we know them. But did we always know? As small children we likely thought very little about our identities. We acted in the world without necessarily pondering our relationship to it. We were autonomous but not fully conscious agents. We received stimuli, we perceived our world and responded to it in a manner that was dictated by a multitude of factors that we could neither comprehend nor were fully aware of.

But that changed at some point, or at a particular point, likely with the hormone rush of puberty. We emerged into a new state of awareness, different from that of a baby, different from that of a pre-pubescent child. First imperceptibly and then all at once, we became people infected with an intense and at times painful self-awareness. We were in a state of being. We noticed our relationships; we started conceptualizing ourselves and our world in new ways.

Past the teen years, we may wonder from time to time— revisit those big questions of who we are and where we were going. In those moments (often when we are feeling lost) we introspect. But even then, we take it on faith that our self-knowledge is sound, that it is an accurate report of objective reality.

So, we carry on, confident that we are who we think we are. We believe that what we are can change and sometimes that it is within our power to effect that change. We reach, we strive, we improve. We even evolve, at least in our own minds.

Through it all, isn't it surprising how little we question the source of our identities? How much we fail to interrogate ourselves until something changes— either trivial or momentous, by our own telling of it? At such crisis points something we believed to have defined us appears to no longer do so. For example, when we lose a job or discover a family secret.

Then, suddenly, we may abandon confidence in our self-knowledge. We may begin to question everything we thought we knew about ourselves, not just the narrow scope of what has changed. Even though such situations present ready openings, that moment may not be the most opportune one to begin a deep self-examination, fraught as it can be.

Instead, it may be preferable to dive in when the seas are calmer, when we can afford the time and mental capacity to burrow below the surface of our self-image, of our life, to discover what lies there, hidden, even to us.

This, then, is the process of deconstruction I am attempting to replicate— a decomposition of the self into its component parts to understand how it is put together and what that means for our sense of who we are. There are myriad ways to do this and over thousands of years of human civilization many of them have been explored; the results charted and reported; accepted and debated.

Still, we don't have all the answers. Thus, it is a path worth re-treading, if only because questions themselves are sources of useful insight. Perhaps we can learn who we think we are and why. Perhaps we can discover what makes us uniquely us. Or we may come away empty-handed, more unsure than when we began. Anything is possible. All roads lead to us.

2

Where We Begin

Before we get into the details, it is worth considering where and when we start to see ourselves as individuals, separate from the world and free to act on it through our choices. As alluded to earlier, puberty certainly seems like the tipping point.

But could it begin at birth or earlier, in the womb? One view is that at such early stages we may be nothing more than automata or "beast machines," as the French philosopher, Rene Descartes believed animals to be. As embryos and fetuses, or even as newborns, we are perhaps not unlike Descartes' conception of animals—mechanical actors who can move but lack subjective experience; beings of flesh and blood whose existence is nevertheless unlit by consciousness.

Alternatively, we are indeed conscious at birth—perhaps even before, but minimally so, if consciousness is apportioned in degrees. And we don't gain a solid, subjective identity— a robust sense of self— until we are much older. If so, does our self-knowledge progress gradually or in fits and starts? Are there defined thresholds—associated with age and/or brain maturity—when we step up from one state to another?

Each of us can only speak for ourselves, only know and communicate our own experiences of these things. More accurately, we can describe what we remember of our experience. The earliest, retrievable memories speak to at least the ability to experience, remember and recall—that is, they hint at some form of consciousness in early childhood. Even then, most of the things we remember from before the age of five or six are likely reconstructed from hearing others talking about us as babies and toddlers rather than from original memories.

Regardless of the veracity of consciousness remembered, how might we have experienced ourselves as children?

The earliest memories are episodic, scraps of film, scratched up not easily affixed in time. The clearest memories are perhaps between the ages of four and six. Everything else is lost. It is as if we didn't exist before then.

For instance, at around age four, I recall having clear ideas about what was interesting or not, about what was worth exploring. I remember undertaking exploration as more than just an autonomic activity that would have been no different from that of an animal scrounging for food or water. But this still feels like a mode of learning— perhaps more sophisticated in scope and scale than that of most other animals— but just the same it was exploration as learning to survive.

It isn't until puberty that a different form of awareness seems to emerge. We start to critically examine ourselves and the world. Began asking why we exist, why anything exists at all. Personally, around that age, I was curious about where all of it came from, what it meant and where it was headed. From that point onwards, it has felt as if my consciousness was being progressively elaborated. Even though my awareness of the experience of having an experience hasn't changed as much since adolescence, my understanding of the experiences, as well as their accumulated weight, does seem to change the quality of each new one.

I can't say what it all means or even if it is something real or only imagined—an illusion that gets deeper and more convincing the longer

I am in it. Either way, psychologically, it feels as if the older I got, the more I settled into a stable identity, more stable than the one I inhabited as a teen or young adult. And there are steppingstones, if not in consciousness, then in self-conception. And perhaps as we enter our senescence, we return to the oblivious-cum-nonchalant state of being that characterizes early childhood, where the idea of self is not something we dwell on. But in between adolescence and middle age, there is a long stretch of time when identity—our very sense of ourselves— is in flux.

So, wherever you yourself are in life, we may as well follow the ancient Delphic maxim, *temet nosce*—human, know thyself.

Consciously or sub-consciously you began as an idea in someone else's mind— either before or soon after you were conceived. At first other people told your story— you smile all the time, you are colicky, you have tantrums; you eat poorly, you sleep well.

At some point a spark ignited or perhaps an embryonic ember became a steady flame, and you began to tell your own story— I like climbing, I am short, I like to wear dresses, I like planes.

But this is merely the construction that we call narrative identity, a biased, self-centred mélange of past, present and future, informed by our environment. To truly understand yourself, you must peel back the layers of the onion that is the story of you—the story that relates the development of your identity as a conscious agent in the world.

Throughout your life you have created and continued to refine a mental representation of yourself. It determines how you see yourself on your own and in relation to others, as well as to the wider world. You reference it when you make decisions. It helps you answer questions such as: What do I think? What kind of person am I? What are my values? Would I do this? Should I?

We tend to see ourselves as mostly stable— a unitary collection of our perceptions, values, desires, heritage and experiences. But we also exist in time, meaning that all those things change as we age. Thus, it is useful to consider Heraclitus of Ephesus' observation that you can never step into the same river twice— one, because the river has changed and two, because you have changed. And just like the river is a stream of water, you are a composition of bones and muscle, flesh, and skin. And you, too, change. You, too, have been changed by your journey through time and space.

Per this view, there are multiple "yous," not a single sustained entity that persists unchanged. You are a process, not an entity. You are impermanent— what the Buddhists call *anicha*. You are a collection of ever-changing memories and thoughts and sensations; those things are not aspects of some core you that is separate; those things are not happening to the real you— they are you!

You are also the story of you, the narrative that you generate based on memories, influenced by genes, and informed by senses— even if that story is not always objective or accurate. It is necessarily distorted, incomplete and inchoate, because all of us present a version of ourselves to the world, largely dictated by how we want to be seen by others, but also by sociocultural forces that attenuate our responses.

Thus, the values we express are the ones we want to be associated with, not necessarily the ones that we practice, as evinced by our actions. We even go so far as to interpret and rationalize our own actions and events to fit a narrative of ourselves— we make the facts fit the story. We also change the story to suit the audience— to impress, please, curry favour, solicit sympathy and so on. We are subject to the self-imposed tyranny of other people's expectations.

And through it all, we cling to our "biodata," not just our professional history, but also the recounting of our personal victories, grudges, qualities, possessions, status, and other characteristics that we believe to be important. It includes the details of where we were born; our parents' and grandparents' names; their professions; our religion; and other bureaucratic details, whose import has long since been forgotten by the very officials who insist on only processing complete forms. But you remember them because they are meaningful to you. These details represent your life.

Yet much of it is history—the inexorable past. History that is quite likely mangled and misrepresented. Nevertheless, therein we are the heroes of our own lives, the captains of our souls. Therein our triumphs are attributable to our own qualities while our failings are attributable to other people's actions. At least that is our penchant—to apply what psychologists call motivated reasoning, post hoc rationalization. It is us doing our own public relations, creating our own spin.

Exactly when we start writing this story is not entirely clear. But one thing is— the story is mediated by memory. The story is written moment by moment, in the eternal present. Yet, that present is layered on sediments laid down over a lifetime. It is also built atop our narrative inheritance, on the stories of our forebears. It is the story of our families / tribes / nations / species— of their histories told and re-told over time, often polished to the shine of myth. It is perhaps an important starting point for our own identities, these memories of our ancestors.

Perhaps your origin story starts there, with your grandparents, the oldest people you are likely to have known reasonably well. Their stories would have been the canvas on which you drew "yourscape," on which you created the palimpsest that is you. It is the masterwork against which you imagine your parents' origins, their idiosyncrasies. If you are lucky, you may have access to records that go further back; or perhaps you are the inheritor of a rich oral tradition that traces your ancestors into the misty, mythic past.

Maybe there are some standouts in that history, persons that acquitted themselves particularly well, who loom large in family lore, with whose memory you may have found yourself competing. Or perhaps there were no heroes, no one who has done anything of historical note, just a steady stream of mediocrity, thankfully lost to time. More likely, whatever their life was like, no one stepped up to tell their stories in any way memorable enough to make themselves resonate through the generations.

For a while and perhaps even now, one story has figured more prominently in your life. A personal history; a tribal myth; a national belief. Whatever form it takes, that one story may have even influenced your own, or other people's telling of it is what has influenced you. But as you grew older, you may have started to explicitly or implicitly accept or reject some of your history.

Then you would have started to revise, to rewrite the story you inherited, even as you kept parts of it, even as you distorted and changed it. This, then, is how we start to define our values, our beliefs and ultimately our whole selves. We either keep or adapt the old; or start fresh, as makes sense to each of us. Whatever the path, the past— both inherited and lived— is inescapable and it continues to define us in ways that we may never completely disentangle or understand.

3

Being Us

The processes that allow us to think, the ones that make us supposedly sentient, are the same ones that allow us to do everything else. Unless you are a dualist— someone who thinks that the stuff of consciousness is special, that mind (and perhaps "soul") is different from the physical brain and its electrochemistry. Of course, talk of the soul or spirit is nothing new. It goes back many thousands of years and across most cultures. This is the notion that the body is merely a machine, a puppet that is un-animated until it is infused with something else that is capable of subjective experience— of consciousness.

It is an idea that has persisted despite scientific advances— in spite of the declining prominence of religion in daily life. Even those that would consider themselves scientific rationalists may have a nagging feeling that the personal, subjective experience of the world is somehow special, in its own category. That "what it is like" to feel something from a first-person perspective cannot be objectively explained. And further, that the necessity of such subjective experience cannot be explained by evolution, since subjective experience does not seem to be necessary for survival. After all, we don't believe that viruses and bacteria have consciousness and they are functional, so human consciousness must be something more, something ineffable— or at least that is the reasoning.

Of course, the very thing that makes us put consciousness on a pedestal also makes it virtually impossible for us to confirm it in microbes or fish or animals or plants or indeed any living thing. We can only devise experiments to help us infer that there is something akin to subjective experience within another living thing. We can never know for certain. In a similar vein, the solipsism that makes us so sure of our own existence and consciousness also makes it impossible for us to confirm that any other human being is conscious in the same manner that we ourselves are. They may be philosophical zombies, identical and able to simulate humans but lacking consciousness.

Yet this unshakable sense of aliveness permeates our individual existence in the world. We think that we are apart from the world, observing it at a distance, even as we inhabit our bodies and are clearly in and of the world. That is why we have myths about the migration of souls from one body to another, from human to animal bodies, existing apart from the body in other realms, of persisting long after our physical forms are gone.

And then there are techno evangelists who believe we will one day be able to upload our consciousness to networked computers (the so-called "cloud") or other storage medium (and perhaps even download it to another body). The goal is for us to survive the inevitable deterioration of the body and even death itself. This would be digital immortality— as long as the computers hosting our consciousness are powered and backed-up.

Such notions are based on what philosophers call dualism— the idea, first advanced by Rene Descartes— that the mind and body are separate entities that interact through some mysterious mechanism. That is, mind acts on matter but is itself not composed of matter. Descartes' famous "cogito, ergo sum" quote arose from this claim— namely, that he could only be sure of his own existence; and that the fact of his existence was indisputable— the one thing he knew for sure. Yet he was convinced that humans, because of their capacity for reason, were unique. That is, a robot, in modern parlance, could never trick a human, no matter how human-like it looked and behaved. In other words, philosophical zombies are impossible.

And he went further, drawing a clear line between humans and animals. To the fellow members of his species, he generously extended the presumption of consciousness, whereas all others he termed "beast machines," without souls. For them there was no subjective experience of colours, sights, sounds, smells, touch— they were simply automatons controlled by a presumably pre-programmed brain.

In the 1990s the philosopher David Chalmers reignited interest in consciousness by declaring it to be the "hard problem." His claim was not that explaining brain function was easy but that there was a clear path to understanding the electro-chemical-mechanics of human existence based on established knowledge and approaches. Consciousness, on the other hand, he contended, was a different class of problem that would not yield to materialist explanations. He popularized a thought experiment by fellow philosopher Thomas Nagel, who in the 1970s, wrote a paper wondering what it was like to be a bat. Nagel suggested that consciousness "feels like something," that being conscious entails knowing "what it is like" to be that being.

Building on Nagel, Chalmers' paper laying out his own views set off a flurry of activity in multiple disciplines, including philosophy, neuroscience and even physics. One of the ideas that has since become prevalent is that consciousness is a fundamental property of the universe, that everything, including quantum fields possess it— and that all matter, whether living or inanimate has a degree of consciousness. This is called panpsychism.

However, panpsychism holds that subjective consciousness only emerges when a certain threshold is crossed. A variation of this view suggests that consciousness is not a property of matter but is its most basic form. That is, consciousness is the ultimate, indivisible thing from which everything else in the universe is made. While this is an intriguing idea, it is almost impossible to verify in any objective way.

Nevertheless, few people would argue that our subjective experience is at the heart of what it is like to be us. It is the defining element of our self-conception. But the question is whether this is something special, something separate from the matter and energy that—to the best of our knowledge—make up the universe.

At the other end of the spectrum are those who believe that consciousness can be completely explained as emerging from biology. Indeed, there is an active scientific research program focused on identifying the so-called neural/neuronal correlates of consciousness (NCC), which seeks to locate the brain regions and perhaps even specific collection of cells that give rise to consciousness.

While many parts of the brain (thalamus, midbrain and pons, etc.) are required for full consciousness, not all of them are necessary to maintain it— as evidenced by people who are either born without parts of the brain (e.g., cerebellum), have suffered damage to certain regions, or have had portions of their brain surgically removed to relieve suffering from disease.

Thus, NCC research is trying to identify the key sub-components that make it possible for us to subjectively experience the world. Ultimately, this research also hopes to determine the specific set of neurochemical processes that cause consciousness. For example, recent evidence shows that electrical stimulation of the claustrum can turn consciousness on and off, but this still does not explain conscious experience.

Other intriguing research provides more clues, including studies of patients who have had their corpus collosum— which is one of the main bundles of nerves that connects the left and right hemispheres of the brain— removed to alleviate seizures. When one eye is covered and they are asked to pick out an object in front of them, such patients can do so without problems. However, they are unable to name the object until the other eye is uncovered. Despite the severed link, they suffer from no other major cognitive deficiencies and live normal lives. Such experiments suggests that consciousness exists independently in each hemisphere.

Some would argue that even if the NCC are found, we would not be able to explain consciousness— especially if consciousness is said to "arise" from the processes or be "caused" by them but remains something different and inexplicable. This is a difficult argument to tackle and may be more a matter of semantics than substance.

For materialists who claim that matter and energy are all there is, the electrochemical processes are what constitutes consciousness— the cause and the effect are one and the same. And here, too, there are multiple theories about why we think we are conscious or why we have subjective experiences at all. Some, like the philosopher Daniel Dennett, suggest that consciousness is to the brain as a graphical user interface is to a computer. It is a pleasing way to access the raw data and capabilities of a complex system. Others, like the neuroscientist Anil Seth, contend that consciousness is an illusion that is necessary to keep us alive. It is simply an experience of the process that acquires and integrates internal and external signals with data that has been previously gathered, stored, and processed to determine the appropriate response or behaviour in any given circumstance.

Another intriguing perspective is that of the philosopher of science, Peter Godfrey-Smith, who posits, based on his extensive study of cephalopods (including octopuses) that consciousness is simply an adaptation to our environment. He thinks consciousness has evolved multiple times, and separately, for example in cephalopods and vertebrates, like humans. Intriguingly, he also suggests that consciousness is nothing more than the experience of our brain processes— they are the feeling of the underlying biology, chemistry and physics.

This, at the meta level, leads us to feel a certain way— as unique, exceptional, unitary. And consciousness aside, many of us, no matter how anxious or under-confident, walk around with a sense of ourselves as exceptional: I wouldn't do that; I am like this, not like that; I think this. We moralize and judge others against our standards— often standards that we believe we meet.

We also separate ourselves from the world but desire to act on it, set specific goals to manipulate it, to have it bend to our will. We believe we are independent agents and that there is clearly something that distinguishes us in a fundamental way. This is implicit, automatic, and possibly necessary for a continuing sense of self. Perhaps without it there can be no ego. And without ego, there may well be no self.

But is it also the greatest self-con of all time? Are we deluded because it is evolutionarily advantageous to feel like we are in control, to believe that we have freedom to act; that the future is not preordained? And for all our self-awareness we cannot access the processes that generate thoughts, that make us believe in our own consciousness.

So exactly how aware are we, really? How conscious could we be if not every detail and nuance of our experience is available to us? In some ways, we seem to have tremendous insight, not only into our own thinking but also into that of others. This is known as "theory of mind." Those who study it believe that humans are capable of impressive feats of inference. That is, we can think about what someone else is thinking about what we are thinking— and so on (what we are thinking they are thinking about what we are thinking about what they are thinking), up to about six orders of intentionality. Animals, too, appear to be capable of multiple orders of intentionality, although it is difficult to determine exactly how many— especially when we can't even confirm each other's consciousness.

Theory of mind is fundamental to our self-understanding as intelligent agents in the world, but we are no closer to understanding what makes it possible. All the same, we can explore theories of consciousness that keep it entirely within the physical realm, without getting into questions about the existence of a soul or even metaphysical ones such as the existence of a distinct mind and body.

The most advanced of these, which are based on a combination of neuroscience and philosophy, suggest that consciousness is nothing more than the brain's model of the body in space and time. Space, because an organization's survival depends on knowing where it is (including where its limbs and other body parts are) so that it can take appropriate action (enabled by the sense of proprioception). Time, because movement is how the body accomplishes its goals—its movement away from pain and threats and towards whatever satisfies its need for sustenance and pleasure.

Thus, the world—or reality—itself becomes a projection of this internal model rather than something that is perceived objectively— as it truly is. Additionally, perception is only a slice of any possible objective reality since our senses cannot detect everything in the environment. We have selective attention, and although we start out in life with the ability to resolve more signals, as we grow older, it is pruned to focus on the signals that are evolutionarily important for our survival. This collection of sensory perceptions of our environment is our "umwelt," or sensorium. This is the world we know.

Importantly, however, even when we can theoretically perceive something in the environment, we do not always do so, partly because of how we have learned to see and partly based on the context of our seeing. That is, we may be more attentive in a potentially dangerous situation that in a casual one. We may be more attuned to the visual and auditory clues that help us detect our child in a crowd than at home.

Consciousness, then, may be a shared hallucination that is if nothing else useful for keeping us alive, as the neuroscientist Anil Seth has suggested. But rather than being a static projection of a fixed model, it seems likely that ours is based on a predictive model that continually calculates and updates probabilities about the state of our umwelt.

In fact, several easy-to-do experiments demonstrate the idea that the brain's expectations determine what it perceives. We can pick out a signal from a noisy audio recording or a figure on the moon when primed, when provided clues about what to hear or what to look for. Similarly, our perception of colour may be influenced by our expectation of things looking darker in shade rather than in direct sunlight, which is the basis of illusions like Adelson's checker-shadow, which appears to show two different colours that are provably the same colour.

This predictive model of reality is continually changing as the brain incorporates the constant flow of new information from the senses, using it to correct prediction errors. And the model can be robust and useful without being complete. We do not need to perceive everything, predict every possibility—just enough to ensure our survival for the time being. Thus, our perception is limited by our attention—a reason why we are highly sensitive to sudden movements or noises but can, and do, easily tune out elements of the world that do not challenge our expectations of it—a constant background din, or activity in the distance.

This brings up the intriguing possibility that we act—we move—before we have fully processed what is happening around us, which may explain many, so-called instinctive reactions. The brain sends instructions to the body based on a pre-existing model and then makes corrections once it has had a chance to process the incoming signals. This is the biological equivalent of moving faster than light— because waiting until all the information has been processed could be fatal.

Experience reinforces the model and helps us to codify behavioural patterns that can be quickly invoked when previously encountered triggers are detected in the environment. The cost of being wrong is not enough to slow down our reactions. This is what allows us to intercept a hockey puck or cricket ball, what lets us anticipate and react to situations faster than the speed of thought.

Indeed, there is evidence that the brain can "autocomplete" the trajectory of a ball or object at twice the speed of its actual movement through space. So, consciousness is not our actual experience but the experience we expect to have. But even that may not be the whole story.

Consider that consciousness may not be one state, but many states and we transition between them throughout the day and night, to say nothing of our entire lives. Sleep, coma, minimally conscious state and wakefulness are all different levels of consciousness wherein we have varying degrees of awareness of our environment and our own thoughts.

Some have pushed the frontiers and have sought to actively cultivate altered states of consciousness through practices such as mindfulness. We could argue that this ancient practice is ultimately about attention without attachment. That is, becoming a witness to our state of mind without partaking in it. While practices like mindfulness and breathwork appear to have measurable benefits, it is questionable whether we can separate ourselves from our consciousness without resorting to dualism and positing a mind that is separate from the physical brain.

Thus, the observer and observed during meditation practice are one and the same, with observations being noted reflexively, shortly after they occur in conscious experience. It seems unlikely that we can perform the extraordinary feat of stepping outside our own minds, to develop and use a meta mind that remains hyper-aware and independent of itself.

Regardless of whether such separation is feasible, it does seem possible to cultivate a higher level of self-awareness whereby we create a pause between thinking and acting—where we stop to consider a thought and make a conscious decision about what to do about it. This may sound simple, but it is immensely powerful—akin to a superpower. It entails developing a state of continual awareness, where even as you are doing or thinking one thing, you are aware both of that thing and of other things that are going on around you. By honing this skill, we may be able to more easily break out of one state of awareness to enter another—but only when and if it needs our attention. This would mean never getting fully absorbed in anything, never being distracted.

One simple example is being primed to respond to a baby even if you are in the midst of an intricate task such as carving a miniature sculpture. You never lose yourself in it, even as you remain focused and immersed in the task at hand. It would seem that the only time this is less possible or altogether impossible is in states of consciousness such as sleep, being under anesthetic or being in a coma—all conditions during which active thought is unavailable, although body awareness is retained to varying degrees, depending on the specific state, and the brain continues the arduous task of keeping us alive, including by waking us up when, say, an unexpected noise signals potential danger.

It would be quite the feat to achieve a state like that, but while it may be easy to divide our attention for some things, it is unlikely to be possible for all things. This is because we have evolved to always to be on alert for specific patterns—again, with the goal of keeping the body in good order—no matter where our attention is focused. And perhaps, with training we can expand that alertness to adjacent patterns or certain ones of our choosing—but not all possible signals or combinations of signals.

Despite being the wonders that they are, our senses have limits, and our brains have only so much power to process what the senses take in. Without the aid of cognitive prostheses, we are unlikely to extend our brains beyond their biological limits.

4

Our Body, Our Self

The "I" that begins life may have synesthesia, no access to language, limited perception, few or no experiences/memories to draw on, and may be both physically and mentally underdeveloped. But that "I" presumably has a sense of itself, a sense of being in a world and even perhaps a notion of being somehow independent of it, of acting on it.

As it matures, that "I" grows more complex, reflexive as it recurses through memory and real-time data from its surroundings— all in the span of milliseconds. Selfhood emerges as layers are laid upon layers, like the millions of years of sediment that record our planet's history and form its identity.

For us, the layers are RNA, DNA, cells, bones, tissues and more. We have trillions of cells in each of our bodies— and everything in us is made of cells (estimates vary and it is difficult to get an exact count). Each of those cells contains the same DNA in its nucleus. This DNA is a tightly wound double-helix, which if unwound would be almost 2 metres long. The DNA contains instructions for making proteins which help to manage bodily processes at the most elementary level. However, only about 2% of our genome appears to encode proteins—

the purpose of the other segments remains largely unknown, although at least some of them may determine how and when proteins are made. But the number of genes we have— either for a specific protein or as a whole organism—does not appear to relate to the complexity of the organism itself, or its ability to perform more sophisticated functions.

Perhaps we began as will—embodied in a gene, trapped in a nucleus, encapsulated within a cell. A gene with the will to ensure its own survival. This is the "selfish" gene hypothesis, formulated by biologist Richard Dawkins, which posits that natural selection and adaptation occur at the gene level to ensure the survival of the host organism and its kin (or similar organisms). In such a view, we are merely a vehicle for the gene's survival; we are not the captains of our ships.

While it may sound reductive, the hypothesis offers an alternative model of the self—one that comprises a multitude, not a singularity. In fact, it could be even stranger. Our genes include DNA from viruses and bacteria that attacked our ancestors (in fact ~8% of the human genome is likely based on DNA from various viruses), and our bodies continue to be hosts to micro-/viro-/myco-biomes—bacteria, viruses, fungi, etc.—that by some estimates may outnumber our own cells.

Of the 75 to 100 trillion cells in the human body, half or more are thought to be cells of other organisms that are as much a part of us as our own cells. It is possible that the actual ratio is much higher— perhaps 1 to 1.5 non-human cells for every human one— although this remains an area of debate. While the study of microbiomes is in its infancy, the organisms that comprise it are thought to influence everything from digestion to behaviour and have been shown to play an active role in the development, reproduction and sociality of several species. It may even influence what traits we pass on to our descendants, by mediating gene expression.

Another sub-discipline that is complementing our understanding of the "microverses" within is epigenetics, which is exploring the many ways that our natural environment (the amount of nutrition we have access to, the predators we are threatened by) as well as our experiences (e.g., relationship with our parents) affect us—not just psychologically, but genetically. It seems possible, though not entirely proven, that some of these factors may determine characteristics passed on to our descendants, with the effects potentially lasting generations.

We may, therefore, not be creatures of our own making. Instead, we are amalgams, derivatives— first, biologically, by incorporating genetic material and information from other organisms; and second, ecologically, in terms of being shaped, from generation to generation, by the actions of other beings and the broader natural environment. What we are and what we will become is not entirely our own doing but inextricably linked to the context of our existence.

Such an understanding starts to further erode the notion of a cohesive self that is independent of its environment, which exists in the world but is not confined by it. For instance, consider findings from the genetic analysis of women who have given birth. Their DNA appears to be altered by that of their offspring, and previous pregnancies affect the genetics of subsequent children. This phenomenon, called fetal microchimerism, results in the child's cells being incorporated into practically every part of the mother's body. The effect of the child's cells on the mother is not entirely clear, though they are thought to be both positive and negative. And while the mother's immune system typically flushes fetal cells after the child is born, by that point they have already been at least partly incorporated into the mother's body and continue to exert varying degrees of influence, including on metabolism, behaviour, and lactation.

On the one hand, the fetal cells may help to regulate the mother's temperature and her ability to provide resources in vivo and after birth. This makes sense evolutionarily, since the child has an interest in maximizing maternal resources for itself. However, doing so could affect the mother's ability to support subsequent pregnancies, setting up what is commonly referred to as the maternal-fetal conflict.

There is also the intriguing possibility that fetal cells that migrate to the mother's brain could have an influence on her behaviour. While research is ongoing, there are indeed structural changes in a mother's brain (thickness of grey matter) that persist for at least several years; and other changes that can last for decades after the baby has been delivered. In essence the child becomes and always remains a part of the mother.

We could interpret examples like these as evidence of genetic determinism. That is, everything we are and everything we do is determined by our genetic code, which would in turn imply that we have no free will— in other words, biology is destiny. Recent claims about the far-reaching effects of genes include the ability to predict, based on genetics and the way it shapes your brain: our political affiliation, our propensity for religious belief, susceptibility to authority, fidelity and a whole host of other characteristics. If true, it would go against many people's sense of agency and identity— their confidence in their ability to control their own actions; and would have tremendous implications for our ideas about personal responsibility and holding people accountable for their choices. It could even undermine the notion of justice itself if we, too, are beast machines.

However, at this point, we don't know enough to conclude that genes determine our entire lives. And even if they did, gene expression is known to be influenced by factors in the external environment, as well as by random fluctuations in the cellular environment.

Perhaps, then, what we have is not unfettered free will but a destiny that is significantly but not wholly mediated by our genetic code. Our genes may largely determine our personalities, physical characteristics, and behaviour. But if we accept the findings of epigenetics, all these aspects of ourselves are also influenced by our life experiences and natural environment (as well as by the experiences of our forebears) — an ongoing dance between nature and nurture to shape what we become. Indeed, depending on the specific environment, the degree to which any one person's characteristics are due to genes or their surroundings could vary widely. Our socioeconomic, natural, and psychological environment, particularly during early childhood, appears to exert a noticeable influence on how we turn out and what aspect of ourselves can be explained by nature vs. nurture.

In twins for example, while there are many superficial similarities, there are also marked differences, especially when those twins have grown up apart, in distinct environments, within which they had divergent experiences. For instance, one twin may get a genetic disease while another one does not. One twin may exhibit behaviour that the other does not. This interplay of environment and genes appears to be evident in a phenomenon known as RNA editing, which is far more common in a type of cephalopod called coleoids, which includes octopuses, squid, and cuttlefish. In most organisms, DNA is transcribed into RNA, which is then used to make proteins that are responsible for most biological functions.

One hypothesis is that the coleoids, which are among the most intelligent species, use RNA editing for rapid— and possibly temporary— adaptation, which avoids reliance on DNA alterations that can take generations and therefore be too late to help the organization survive an immediate peril. This RNA editing may happen in response to environmental changes such as temperature fluctuations in the ocean. Regardless of the trigger, most of the editing happens in the neurons of the brain, with a high degree of variance among neurons.

To a far lesser extent, RNA editing is also found in humans and other forms of life. Although the reason for its existence remains unclear and it challenges the so-called central dogma of biology (DNA makes RNA makes proteins), a leading theory suggests that such editing enabled repair and is thus beneficial to the organism. Nevertheless, genes and protein production are influenced by the environment; our bodies, and therefore our selves, are not static in the face of a changing environment— they must adapt or perish.

Again, we are reminded that any thought of having evolved beyond the confines of the natural world is fantasy. The very notion of an individual is suspect. Beyond our role and place in society and the environment, we are also part of a larger ecosystem, including not just the earth but also the sun, moon and beyond. The solar wind affects weather, the moon influences tides, neutrinos from the depths of space are continually passing through us and may influence us in ways we haven't yet understood.

The unity of our selves is also an illusion; as is the narrative of the individual as sovereign, fighting to maintain his or her liberty, values and authenticity in society and within the natural environment. Even aside from what is physically inside or outside of us, our sense of self is surprisingly fragile and easily disrupted. This has become evident through the study of diseases such as schizophrenia, which appears— at least in part—to be an inability to separate oneself from others. It could be one of the reasons that schizophrenics attribute intentions and sometimes even their own actions to others (an altered sense of agency). The afflicted don't always seem to have a cohesive sense of self or a clear notion of where they stop, and others begin.

Similarly, those who suffer from body dysphoria disown parts of their own body—from not recognizing an arm or a leg to doubting their own existence (as in some forms of Cotard's syndrome wherein people believe that they are dead). Some have even gone so far as to have a limb amputated and replaced with a prosthetic. This sense of body ownership can be readily replicated with the so-called rubber hand illusion experiment, subjects of which begin to think that a rubber hand is their own. They gradually start to "feel" touches or strokes on the fake hand as being experienced on their own, real, hand, which is hidden from view. Similar effects can be achieved by donning virtual reality glasses, which modify or swap one's own body with other forms to alter our sense of body ownership.

What this suggests is that our sense of self (which is posited to be an integration of sensory information, thoughts and emotions) is also embedded in the body and movement. That is, I move, therefore I am. Indeed, neuroscientists have suggested that emotions are felt in the body before we become aware of them—from sweaty palms to butterflies in the stomach, brain regions such as the basal ganglia. The latter mediates learning and intuition and plays a major role in communicating feelings that well up in response to external events. Psychology, then, is physiology; body is mind.

Contrary to traditional thinking that the brain evolved as a central command centre, cognition is now thought to be a kludge, a rag-tag assembly of systems that developed to address different needs— most of them related to the need to manage the body— that have subsequently become linked. However, even now, the central processing unit is too slow for some purposes— processing needs to happen closer to the edge, closer to the muscles that control movement. This raises the intriguing possibility that our thoughts are generated by movement rather than the other way around. What's more, we can have a sense of self that is based on being in our bodies (embodiment), which may be distinct from the sense of self we have when we can use language to articulate thoughts about ourselves, to create a narrative self. Babies, for instance, and perhaps many animals have an embodied sense of self even if they lack a narrative one.

More fundamental than this evidence of the brain-body relationship is evidence that the cells that make up our body are quite distinct from each other. As much as we tend to think of each cell as having the same DNA— making us who we are— it is now evident that this is not entirely so. Different genes may be turned on in different cells—almost making it seem like they came from different people. This is called mosaicism and may be far more common than once thought. And it is not just a feature of cancer cells, which mutate and then spread, becoming different from the cells that surround them.

Depending on when they happen— early or later in life— will determine how much mutations spread and to what extent they cause diseases. Regardless, mutation— and therefore mosaicism— is common. Cells themselves have lineages, which carry their legacy of mutations down through the generations. We are then, not one thing, but many. Even different parts of our body are genetically distinct.

Finally, there is the question of what is innate. That is, what traits we are born with and what we develop based on how, where and with whom we grow up. While this has traditionally been framed as the nurture vs. nature debate, newer research suggests that random events play a significant role in gene expression and biological development, as well as throughout our lives. Specifically, the conditions within and around a cell at a particular moment in time—when DNA is being translated to RNA, which is turn is translated into proteins—can affect both the type, quality and quantity of proteins produced. This then influences the functioning of the entire organism, including its behaviour.

And despite tremendous advances in understanding how genes influence our health and behaviour, and even some success in manipulating them at an embryonic stage (using techniques such as CRISPR that can snip and replace DNA segments), much remains unknown. One area of debate that is also politically charged is that of genetic differences between men and women or among so-called races and how this may manifest in areas such as intelligence and the ability to do certain kinds of work. Understanding this scientifically requires a more fine-grained exploration of minute gene variations and their domino effects.

Even though humans share ~99.9% of our DNA with each other, the

0.01% difference accounts for significant diversity. The question is how that variation alters our selves in ways that may be beyond our ability to influence. For instance, DNA determines what proteins are made and the specific structure of those proteins (proteins "fold" in different ways and the resulting three-dimensional structure can determine the function of the protein) and other factors, including the cellular environment can determine what proteins are produced and how much is produced when.

Proteins are used to build cellular structures, which in turn determine brain structure. And brain structure, including both the size of various regions and, perhaps more importantly, the connections among them, determine brain function and therefore our abilities, both intellectual and physical.

However, the influence of the size of various brain regions or even the number of neurons in them on observable abilities has long been debated. For instance, both elephants and whales have larger brains than humans but do not, on the whole, appear to be more cognitively capable than us. Within our own brains, the cerebellum at the back of the brain, which handles a lot of unconscious bodily processes has more neurons than the cerebrum, which is responsible for higher order functions.

Thus, while neither brain size nor number of neurons seems to correlate directly with intelligence, these factors may yet play a role in other capabilities that we have not been able to pinpoint. For now, the interconnectivity of neurons seems to be a bigger factor in human intelligence when compared to other animals. Research into the "connectome," as this has become known, has already identified specific sets of connections, described as networks, that seem to be involved in active thinking, mind wandering, consciousness and other functions.

Despite everything we know now, most of us continue to believe that biology is not destiny. That conclusion is premised on the existence of free will, that is, on humans having the ability to decouple cause from effect. If we don't want to accept biological determinism, we also have to assume that our behaviour, which is at least initially a function of our brain structure (e.g., neural connectivity determined by proteins manufactured based on instructions in DNA), can in turn modify brain structure to change our abilities. That is, we have to assume that modifying behaviour will change the brain, much as exercise changes muscle fibres.

To some extent there is evidence for this so-called brain plasticity— the ability of the brain to adapt— just as muscles adapt to the strain caused by training. Perhaps most famously, London cab drivers who are famed for their intricate knowledge of the city's streets, appear to have on average, larger hippocampi. Similarly, women who were historically believed to have poorer spatial skills, have been shown to match men's spatial abilities with enough training (for instance, by spending a lot of time playing video games).

But the confounding factor may be that even our choices, our decisions to modify our behaviour or change our environment, may themselves be determined by our genes and the random variations that these are subject to. All of which could mean that we ultimately have no control, no free will, no ability to actually change things, only be changed by them in ways that we cannot actually influence.

Let us, then, return to the controversial question of genes, gender, race, and other differences among humans. To what extent do these manifest as differences in abilities? At an individual level, there seems to be little debate that we are often different from each other in temperament and capabilities, including intellect, athletic abilities and so on. And by most measures, the differences between individuals within various groups (men, women, ethnic groups, etc.) are greater than the differences between the groups themselves.

Thus, race may be a social rather than a biological construct and any differences between men and women, beyond obvious anatomical ones, could be explained by their different social experiences, from childhood onwards. For instance, by how girls are raised to care for others while boys are taught to be strong, affectless men. However, here, too, the debate goes on with some suggesting that millions of years of evolution has differentiated the sexes and to a lesser extent, tens-of-thousands of years of isolation have similarly led to clear and distinct genetic differences between people in Europe, Australia, East Asia and Africa, for instance.

It is likely that this picture, too, may change. We may discover that even if those variations exist, they do not lead to meaningful differences in abilities or behaviour. Or that even if they do, it is possible to practically erase some such differences through learning or physical training. And there is no telling what scientific evidence will emerge or what political and cultural consensus we can reach about how and even whether to classify humans in some manner— race, ethnicity etc.— for scientific purposes. And finally, there is the question of how all of this will play out in a world where increased global migration is leading to an unprecedented intermingling of cultures and genes.

But we don't have to wait to decide how this knowledge should affect our perception of both ourselves and those around us. We could simply agree to recognize whatever differences do exist without using them as a basis of discrimination—for treating others as lesser than us. We may never come to see each other—within or across groups—as truly equal but we can still practice compassion.

Or can we?

For, we should acknowledge that not everyone will agree, that our biases will likely persist despite any evidence to the contrary. We can reasonably predict that there will continue to be differences in opinion and thus ongoing discrimination and maltreatment. After all, even as we talk about treating animals humanely (essentially suggesting that we should treat them something like how we treat each other) we do still sometimes treat our pets better than we treat other people.

Perhaps, we cannot fully escape tribalism because, as much as we may not want to admit, it offers some evolutionary advantage. That is, acting as if blood is thicker than water, increases our chances of propagating our own genes rather than that of others who may compete for the same resources.

All the same, we can always try. This much is clearly within our abilities— to listen to the better angels of our nature, as the 16th American president, Abraham Lincoln, put it in his first inaugural address. When we think of ourselves and whatever perceived abilities or shortcomings we have, we can try to not judge ourselves too loftily or harshly by them.

Neither should we feel limited or shortchanged by what we cannot do, by how unfavourably we may compare with others; or indeed give in to fatalism, even if there is no free will. For the sake of our sanity, it helps to act as if we do have the freedom to choose and besides, we may not be able to do otherwise, no matter how much we try.

5

Made by Memory

It is curious that so much of our sense of self is mediated by memory, by remembered experiences. Its reflexive nature and its continuity through time makes one form of consciousness possible. That is because being happens in time, and without memory there would be no self to refer back to or experience the changes in feelings and perceptions over time— there would be nothing to generate a sense of continuity.

Some scientists also argue that memory and perception are intertwined, in part because data collected through the senses and processed by the brain is held in memory and retrieved or reconstructed to aid cognition. Conscious experience—at least the kind that humans experience—may not be possible without the interplay between memory, perception, and possibly language.

Despite being a requirement for self-continuity, cases from the medical literature highlight the persistent puzzle of memory's role. Consider British musicologist and accomplished musician, Clive Wearing, who contracted a herpes simplex infection in the mid-1980s, when he was about forty-seven years old. The virus attacked his brain and central nervous system, causing damage to his hippocampus. The resulting amnesia left him with a short-term memory that was limited to a mere 7 seconds.

Even though long-term memories were also affected, he could still play the piano and recall certain incidents from his childhood, especially ones involving numbers, as well being able to remember his relationships to other people. It was as if certain, skill-related memories were preserved while more experiential ones were lost.

His perception and ability to interpret sensory data were not affected. Nor were his skills or knowledge. All of that had been preserved, even if he wasn't directly aware of it. But now his existence was literally moment to moment— he was stuck in a permanent present.

Every few seconds, his awareness would reset, and it was as if he had just woken up albeit with the benefits of having been conscious for a lifetime. Further, he could no longer form any new memories and, since his amnesia was retrograde, most of his awareness of his own past had also been wiped out.

In effect, Wearing no longer had a continuous narrative of his own existence. Without the ability to remember what happened more than a few moments ago, he had to rely on older memories for some sense of self, supplemented with glimpses of current experience that may have augmented his understanding of himself at any given moment.

While this condition created numerous practical difficulties for Wearing and his family, he still sees himself as an individual in the world although often a confused one. In captured film footage, he appears aware that something is amiss but is not sure about exactly what. Even though he knows that he exists, he reportedly doesn't feel entirely conscious or completely alive. Every time his memory was wiped, it was like emerging into consciousness for the first time, except it came with a variety of built-in capabilities— abilities that a newborn baby wouldn't have.

Nevertheless, for him it must be bewildering, frightening, and enraging . . . a Sisyphean cycle that he doesn't even know he is in, wherein existence is ephemeral yet eternal— perhaps the very definition of a living hell, except that he is somewhat oblivious to it.

Indeed, excerpts from his diary entries, published by his wife, Deborah Wearing in "Forever Today: A Memoir of Love and Amnesia," reveal Clive's anguish. One entry begins with him noting that he is awake for the first time and then continues at approximately half-our intervals with further declarations of: "Now I am awake!" And at the top of the page, he's written in big letters "I DO live!!!!"

What kind of existence does this represent? What kind of self does Wearing have, now, being someone who is only episodically aware of his existence but is unable to do anything with that knowledge because of the inability to augment his narrative identity? Was Wearing truly conscious or was he an automaton constantly switched on and off, unable to even control the switch? The one light in his life, the one person he recognized through all this, was his wife. And as he repeatedly fell into depression, she remained a source of comfort, as specifically noted in his diaries. And over time, he seemed to compensate by inventing his past, using his imagination so that it was difficult for anyone who didn't know him to tell whether whatever he was relating was a true recollection or inventions of his altered mind.

Another remarkable discovery that came decades after the original diagnosis was that he could form new memories after all, although they were implicit, rather than explicit ones. He couldn't tell anyone what he had learned but his actions suggested that he was forming enough memories to adapt to new situations, to infer where he was, what he had been doing, how to do something, etc.

Essentially, his body was able to remember, even if his conscious mind did not. Some of these implicit memories may also have been emotional ones, like the ones that allowed him to recognize his wife and children. These are embedded and encoded somewhere in the recesses of our brains and affect our behaviours in ways that we are mostly unaware of. It is what we come to think of as instinct—our automatic, programmed responses to stimuli, quite visible in those that have been through trauma. We can't always explain why we reacted as we did, but obviously somewhere the lessons are encoded and readily retrieved. There is no need for them to inform our consciousness.

Research by neuroscientist Charan Ranganath has shown that different brain areas and networks are involved in short-term (working) memory vs long-term memory. He implicates two cortical systems, the dorsolateral prefrontal cortex (DLPFC) for working memory and the ventromedial prefrontal cortex (VMPFC) for long-term memory. Indeed, the prefrontal cortex is now thought to play an integral to memory formation, as opposed to just the hippocampus. And this is especially true for emotion-related memories.

It is curious that although conscious memories are the basis of consciousness, unconscious ones are essential for many of the things that we associate with ourselves, including skills and abilities. In fact, conscious thought disrupts the learned action pattern, slowing us down, making us prone to mistakes. Consider that we sometimes need to stop paying attention to our hands on a keyboard when typing a password to enter it correctly or to correctly play a musical piece on an instrument. It is the position of the fingers and the order of movements we remember, not the individual letters of the password or sequence of notes.

Wearing's case raises the question of how important memories are to our ongoing self-image. Is it possible that experiences may permanently change our brains to differing extents, and after a certain point memories are not required to maintain that aspect of the self? Is this not the essence of learning, after all? But learning goes beyond knowledge and skills to also inform self-conception, however vague and undefined—perhaps it also helps us to remember that we are us.

This separation between conscious and unconscious memory also suggests that how we present to others and how we know ourselves may be different. Indeed, we may not have access to how we know something, as starkly exposed by Wearing's condition, and may not be able to explain it to someone else, yet we clearly have an ability that other people perceive. Their judgment of these qualities informs their view of us, but perhaps in some cases, does not inform our view of ourselves.

Obviously, from a sociological and psychological perspective this goes beyond just conscious vs. unconscious memory. Other people's conception of us, and ours of them, is subject to our respective experiences and traits and is unavoidably subjective.

Now, contrast Wearing's case with that of people like Jill Price, sometimes called the "woman who can't forget." Those who have tested her suggest that she does indeed remember everything— not necessarily from birth but from a specific point in time. Price and people like her are said to have highly superior autobiographical memory (HASM); sometimes referred to as hyperthymesia. How is Price's sense of herself different from everyone else's? When you can presumably remember every minutia of your existence, how does it change how you experience it.

Forgetting is as important as remembering for most of us. If we remembered every slight, every failure, every humiliation, how could we move on? How can we overcome anything? How can we reinvent ourselves?

Neuroscientists Ronal Davis and Yi Zhong have even identified specific brain cells that they have labeled "forgetting cells," which appear to be responsible for degrading the engrams in memory cells through a defined process controlled by neurotransmitters and specific proteins.

Whether retained or lost and retrieved, memory is fallible; it can be manipulated—both intentionally and unintentionally. Take, for instance, stories related by both American journalist Brian Williams—of being under attack while flying in a helicopter during the Iraq war—and former U.S. First Lady and Secretary of State, Hillary Clinton, who recalled coming under sniper fire while on a visit to Bosnia in the 1990s. Both reports were proven to be false—yet, given how easily disprovable they were, there appears to have been no intent to mislead—simply misremembrance or conflation of what they had heard or seen with their own remembered experience.

This may even extend to some instances of plagiarism by writers, where something they had read or heard appears to be their own thought. Yet, we judge such people harshly—we place high value on originality without fully understanding what it is—or its limitations. Is any thought or idea wholly original, independent of everything that has gone before?

Beyond the realm of ideas, the fallibility of memories affects many other aspects of our existence, including relationships and justice. The psychologist, Elizabeth Loftus, has studied this topic extensively and has concluded that human memory is malleable and that we can be tricked into remembering things that never happened. Indeed, other researchers support this theory with studies that show how we tend to remember our own actions, predictions, and behaviour more favourably and nobly than those of others. We reinterpret ourselves based on the present situation so that we can remain the heroes of our own lives.

Similarly, experiments using transcranial stimulation (TMS), wherein a subject's brain is placed in a magnetic field, has shown that it is possible to change someone's confidence in a recollection, without altering the memory itself. TMS can affect the memory's vividness, detail, and even the subjective experience of it. It appears to do so by disrupting the angular gyrus, a region in the parietal lobe of the brain, which interacts with the hippocampus and sensory processing regions to integrate audio visual data.

Despite this, some studies have suggested that emotionally charged memories are less subject to distortion— and are indeed remembered with additional context— than more neutral memories, such as where you left your keys. There appears to be a part of the brain that seems to be responsible for so-called "reality monitoring," that is, distinguishing between whether I imagined turning the stove off or remember doing so. This is the paracingulate sulcus length, a fold in the brain, which when absent impairs the ability to separate what is real from what is imagined. It is thought to possibly play a role in schizophrenia, whose sufferers have difficulty separating hallucinations from reality.

We are our memories—as flawed as they are. Our values are defined by our lived and remembered experiences. In Japanese director Kurosawa's classic 1950s film, Rashomon, different people (a Samurai, his wife, a bandit, a woodcutter, and a priest) recount different, conflicting versions of the events that led up to the death of the Samurai. It is unclear who is telling the truth, or even if they believe they are telling the truth. We are left to decide whose memory is the most accurate, which version is most correlated with reality.

This suggests that memory is not a recording but rather a reconstruction— that may be different each time it is recalled. Indeed, memories are not static—rather they change over time, as new experience is overlaid on old, and as they are recalled, reconstructed, and re-remembered. Additionally, the brain circuits that encode a memory (within the hippocampus) are not the same as the ones that recall it (within the subiculum of the hippocampus). This is possibly because there is benefit in encoding a new memory while at the same time using an old one to inform an ongoing situation— possibly a dangerous one. Ultimately, what we consider experience, or remembrance may be nothing more than a pattern of neuron activation associated with a given event or concept, or a schema.

Despite this, we are overconfident about our own recollections while doubting that of others. In fact, there is evidence that witness accounts of an event often vary widely from video recordings of the same events, which (deepfakes generated by artificial intelligence aside) arguably provide the most objective record of what happened. These cognitive distortions may arise because of our tendency to prefer narrative coherence— that is, maintain a sense of identity that we have developed over time, with remembrances that most closely reflect the story we have come to tell about ourselves. This is apparent in amnesiacs and patients with various forms of dementia, who hold to very specific memories, mostly from the formative years when they were solidifying their identities as persons in the world.

Flawed though it is, memory can also be a powerful tool. Ancient memorization techniques demonstrate the extent to which we can train ourselves to remember accurately—at least for a short period. The so-called "method of loci" or "memory palace" has been used to recall the exact order of a full deck of cards or up to 70,000 digits of Pi. The method entails placing the thing to be remembered in a physical space—often a familiar one like your home, and frequently, in absurd or surprising configurations (e.g., your keys being held in the teeth of a dragon perched on your sideboard). Recall involves walking through the different rooms in the space, which triggers a memory of the image associated with each location.

In the end, perhaps we are not our memories, though what we remember of our past, or at least a version of it, informs our day-to-day existence. Clearly, we can be alive without a complete memory of ourselves, as evidenced by amnesiacs, and those who suffer from dementia and other diseases that impact the ability to recall not only events but also information about ourselves and the world around us. And even if our memories are intact, parts of ourselves are hidden from us, in the form of implicit memories that are not accessible to the conscious mind. Knowing all this, we could conclude that we are the result of a process, not a conscious construction of experiences but rather the result of both deliberate and automatic events— both inside us and out in the world.

6

Thinking Beings

To adapt is to survive and it is quite likely in service of survival that we have the cognitive abilities that we do today. It is also for that reason that we are born with immature brains that make it difficult for us to survive for long without caregivers— unlike other animals that live and thrive independently right at, or shortly after, birth. But the advantage of the immature brain is that it has room to grow, to form new neuronal connections and complex structures that are better suited to, and reflect, the environs in which it finds itself. Our brains grow as we grow, and that improves the probability of survival.

Those brains are capable of thought and subjective conscious experience. The nature of thought itself is mysterious at first blush. However, it seems likely that it is nothing more than the brain's processing of memory and sensory data, although not everyone would agree with such a reductive conclusion. They would argue that thought is a property of consciousness, the origins and nature of which remain a mystery, perhaps even mystical.

However, if we take the first view, it is also conceivable— as psychologists like Nick Chater have argued— that there is no such thing as unconscious thought— just a set of processes that provide input into conscious thought. This means that all thoughts are conscious ones, although we may access memories and process sensory input in the background.

The German physician and polymath, Herman von Helmholtz, was among the first to posit that our perception, and therefore, thought is based on inferences about the world. He understood that all the data from our senses is insufficient to provide a complete picture of the world or to help create a comprehensive and representative model of it in our brains.

Thus, our brains reconstruct the world using raw data, pre-processed information from memory and inferences to fill in the gaps. Based on this amalgamated dataset, we set out to generate thoughts about the world and about ourselves— so-called reflexive thoughts that help us form an impression of who we are, our opinions of ourselves and what is outside of us. We likely also rely on some innate models of the world, built over hundreds-of-thousands, or perhaps millions of years of evolution that prepare us to survive in the world— for example, to recognize a caregiver's face or respond to certain sounds, smells and other signals. What those models are, and to what extent our understanding is pre-configured, remains a topic of speculation and research but all available evidence suggests that these models do exist.

Even with models, we would not be able to effectively navigate and thrive in the environment without thought of some kind, although it probably does not always have to involve language. For instance, we are aware of a need to relieve ourselves, quench thirst or scratch an itch, and will often automatically take the action necessary to respond to that condition, without articulating it through language. Whether this is a form of unconscious thinking or no thought at all is debatable, but it is a form of cognition— of the animal responding to internal or external state changes.

While dualists may disagree, thought has a biological basis and specific electrochemical characteristics. It can, however, take many forms, from goal-directed, active thinking, to disassociated daydreaming; from imagination to remembrance; or from reflection to barely conscious information processing. All this from an organ that constitutes about 2% of our body mass but accounts for about 20% of a human body's total energy consumption—about 20 watts of electrical energy. Comparing these power requirements with the computing potential of a human brain demonstrates how far silicon-based computers must go to match the complexity and efficiency that evolution has accomplished, without really trying.

At its most fundamental level, the brain is nothing more that a series of interconnected electric circuits. And experiments by Robert M.G. Reinhart and colleagues have shown that by manipulating or recreating the signal patterns that traverse those circuits, it is possible to treat a variety of psychological conditions—from depression to obsessive compulsive disorder. In some ways, this is a more sophisticated form of electro-shock therapy that was once widely used for similar conditions.

The modern method tests a variety of low-voltage electrical stimuli (applied to different parts of the brain at different frequencies) to understand how it affects a particular condition or person. Based on the results, an electrode inserted into the patient's brain can be programmed to replay the most beneficial patterns. This has been shown to instantaneously improve mood or significantly mitigate several types of undesirable behaviours. With the emergence of artificial intelligence (AI), experiments to personalize deep brain stimulation (DBS) using AI trained on patient's own neural data, has shown to be more effective than traditional DBS.

Such findings suggest that ultimately everything may be reducible to neuron firing frequencies, which however are unique to each individual. By determining the patterns that work for each person, it may be possible to address many, if not all, psychological conditions—and to do so in a much more effective manner than is possible with medication or therapy. Nonetheless, given the variation of these neural firing patterns across individuals, it is possible that such stimulation alone will not be enough to effectively treat some people.

The brain is packed with about 100 billion neurons, which are the equivalent of wires in a home's electrical circuit or the transistors in a microchip. However, the number of neurons themselves seem to be less important to cognitive capability than the connections among them. These connections are much more numerous in babies but are typically pruned as we grow. This so-called synaptic pruning appears to be a way of keeping the brain operating as efficiently as possible by removing connections that are no longer needed. It is also the basis of brain plasticity—that is, the brain's ability to adapt to a changing environment, including, in extreme cases, recovering from damage to parts of the organ itself or to any of the senses.

Within the brain, different major regions have different numbers of neurons, with the cerebellum (in the back of the head and largely responsible for coordinating movement) having more than the cerebrum (at the front and top of the head). And then there are the networks, bundles of nerve fibre that connect different parts of the brain, with each bundle activating to enable specific functions.

Diseases, physical damage, and external stimulation have shown that the brain's internal communication system, which is facilitated by these connections, can be disrupted. In fact, studies of people experiencing such states are being used to create a comprehensive map of the brain's interconnections.

Despite this trove of research findings, scientists still don't have answers to key questions such as the origin, end-to-end mechanism and content of thoughts. This is even though, using functional magnetic resonance imaging (fMRI), they can track brain "activations" when thinking about different things or performing various tasks. These activations are inferred by monitoring the flow of oxygen to different parts of the brain during an fMRI scan, under the assumption that blood flow and resulting oxygenation are indicative of elevated activity. This remains, at best a low-resolution tool for understanding what is really going on in our brains when we undertake various tasks.

Nevertheless, what is already known from fMRI and other brain studies is fascinating. For instance, the part of the brain where the temporal and parietal lobes meet— the so called temporoparietal junction or TPJ— can be stimulated to generate out of body experiences, feelings of other presences, images of ghostly figures and altered perceptions of oneself and others. There is also a phenomenon known as autoscopy, which makes people hallucinate their doppelgänger and see them acting independently of themselves, sometimes even leading to confusion about which one of the two selves is real. These altered states are theorized to arise from a disconnect between the brain's model of the body (the "self") relative to its environment and the body's actual location in space.

The TPJ has also been implicated in moral decision-making and damage to it can impair someone's ability to distinguish right from wrong, even when their knowledge about a particular system of ethics remains intact. Further, the TPJ seems to play a role in overall social cognition, that is, in how we relate to others; as well as our own sense of agency and our ability to focus. Later, we will see how it plays a role in forming our identity.

At a more fundamental level, small changes in the chemical composition of the brain— often manifesting as varying hormone or neurotransmitter levels— can influence social bonding (mother/child, romantic partners, etc.). Specifically, changes in oxytocin levels can affect someone's ability to distinguish between who is part of their group and who is not, or even the degree of favouritism exhibited among in-group members or prejudice towards out-group members.

Beyond the chemicals already in our bodies, those we ingest can have a noticeable impact on our thinking and overall cognition. For example, psychedelics such as psilocybin found in so-called magic mushrooms and drugs like LSD seem to alter perception, thinking and overall consciousness. In the 1990s and into the 21st century, following a period when it was banned or frowned upon, scientists re-started research programs into the impact of psychedelics on the brain, including its potential as a treatment for depression, addiction and other mental illnesses.

Similarly, ayahuasca (a brew made from a combination of different plants, which had long used in the ceremonies of indigenous Americans) also has psychoactive effects and possible medicinal properties. Intriguingly, early studies suggest that their effects— e.g., feelings of connectedness and calm— are long-lasting, persisting well after the drug itself has been cleared from the body.

So, if there is any conclusion we can draw about how we think, it is that thinking itself is influenced by a multitude of factors, both intrinsic and extrinsic to ourselves. And as much as we want to believe that we are the ones thinking our thoughts; and as much as we want to be the controller of those thoughts, this may not be the case, even in the absence of a mental disorder. Everything from small chemical changes and electrical interference to defects and disease can affect what and how we think, not to mention evolutionary adaptations and our current social, natural, and cosmic environment.

Consider, for instance, that the first mass shooting in the United States, which occurred in 1966 at the University of Texas was attributed to an undetected brain tumour that led the gunman to develop hallucinations and paranoia. More recently, we have learned that even beyond such obvious influences, bacteria in our diet can change the signals sent to our brains— possibly impacting our moods and perhaps other brain activity as well.

Among the most interesting findings in human cognition have come from research into split brain consciousness and attempts to measure neuron activity that precedes conscious awareness of an action. Both these types of experiments were initially conducted in the 1980s. The split-brain ones, led by cognitive neuroscientist Michael Gazzaniga, involved showing different images to each eye. The left hemisphere (which is mostly responsible for language) is shown one image via the right eye; and the right hemisphere sees another image projected on the retina of the left eye.

The patients who participated in these experiments had had their corpus callosum severed as a treatment for frequent epileptic seizures. The corpus callosum is a thick bundle of nerve fibres that connects the left and right hemispheres of the brain. When it is severed, there seems to be little noticeable difference in the patient's cognitive function. However, in the split-brain experiments where an image projected to only one eye was available to only one hemisphere (in this case an image on the left retina that was only available to the right hemisphere), patients had no way of describing what they saw. Nonetheless, they could pick up the object when asked to do so.

The other curious finding was about how these patients rationalized their actions. For instance, in one experiment the right hemisphere (left retina) was shown a snow scene, while the left hemisphere (right retina) was shown a picture of a chicken's foot. Visible in front of the patients and available to both eyes / hemispheres was a set of cards with pictures associated with each of the two images projected onto the patient's retinas.

When asked to pick up the related card, each hand picked up the appropriate card (in the case of the snow scene shown to the right hemisphere, the left hand pointed to the shovel. But when asked to explain this choice, the patient was initially at a loss, but quickly made up a story about having to use the shovel in the chicken shed.

This is a classic and oft-cited example about the brain's tendency to make up stories, to interpret what it perceives so that it makes sense and fits into an established narrative— in this case the left hemisphere had seen the chicken foot and the patient could see and was aware of having picked up the shovel card, but not why they had done it, since the snow scene was not available to the left hemisphere's language processing network.

In a similar fashion, the neuroscientist Benjamin Libet, building on experiments by German scientists in the 1960s, measured EEG signals from participants' brains and then asked them to flex either wrist at their discretion throughout the experiment. Subjects were also asked to note the time at which they decided to flex their wrist by referring to a special clock that was visible to them.

The results suggested that the subjects' brains were getting ready to flex the arm muscle as much as 500 milliseconds before their muscles actually flexed, whereas the participants reported making the decision only about 150 milliseconds before their muscle flexed. Using this "readiness potential," as Libet termed it, he could predict when a participant was going to flex their wrist, often before they were consciously aware of making the decision to do so— after correcting for errors, he estimated that the signal to flex was initiated about 400 milliseconds before the volition entered conscious awareness.

Starting in 2008, John-Dylan Haynes re-created Libet's experiment, but this time using functional magnetic resonance imaging (fMRI). He concluded that the prediction window could be extended to as much as 10 seconds before the subject's reported conscious awareness of a decision to push a button.

However, in 2012, neuroscientist Aaron Schruger and colleagues noted that the readiness potential observed by Libet, and others may be nothing more than spontaneous and periodic waves of neuronal activity that occur even in the absence of specific physical actions. Their experiment seemed to indicate that this activity existed regardless of a subject's volition.

Regardless of whether such experiments prove or disprove the existence of free will, it suggests that decision making may not be entirely conscious. That is, even if there isn't a specific sequence of neuron firings prior to conscious awareness of volition, there is still a question of whether our actions are somehow influenced by these naturally occurring "brain waves."

Thus, it is possible that our reported reasons for making a particular decision may not be what we say they are. It hints at an unconscious cascade that begins elsewhere in the brain before it enters our conscious awareness. In effect, we may not be in as much control of our own thoughts and actions. All this to say that thoughts alone do not make the person, especially when those thoughts are not the product of free will but the children of genes, memory, environment and so much more.

7

Bugs for Brains

To be human is to fail. Unfailingly. This is obvious from observing oneself and others engaging in routine activities, making everyday decisions—without having to resort to religious concepts such as original sin.

As much as we have developed reason as a tool to reduce our fallibility and despite its success in the practice of science, we continue to be prone to error— to an incorrigible irrationality. Such irrationality is often the result of rapid decision making— what the psychologists Daniel Kahneman and Amos Tversky called System 1 thinking—the automatic, intuitive responses honed by habit and implicit assumptions. These become brain heuristics or more commonly, biases. At least since the Enlightenment, we have assumed that humans are creatures of reason, which is what sets us apart from animals; that emotions lead us astray; and that we always act in our own interest.

But this has since been upended by behavioural economics, pioneered by Kahneman and Tversky, which introduced the idea of two modes of thinking—System 1 and System 2—that guide human decision making. The pair suggested that System 1 engages first and automatically—and that, often, a pause is required to engage System 2, the more deliberate, analytical part of the brain. Nevertheless, both Systems 1 and 2 can lead to either rational or irrational thinking. And there is ongoing debate about whether prolonged engagement in System 2 activities could lead to a takeover by System 1—something called "ego depletion." This is based on the idea that there is a limit to our capacity for sustained deep, analytical thinking, and overtaxing it may weaken our ability to act deliberately.

Regardless of the specific type of thinking, we are notoriously bad at making predictions— unless we take a cold, analytical approach based on reliable data within a domain that is relatively stable. For instance, while it has had its issues, polling is typically considered to be a robust basis for making predictions about elections. However, it is still dependent on people telling the truth, selecting a truly representative sample and on questions being constructed in a manner that does not bias the results. Similarly, chess grandmasters can make good predictions by memorizing typical arrangements of pieces on a board. They can then quickly recognize patterns on the board, assess the state of play and predict their opponent's likely moves from that point forward. The latter is a limited form of prediction based on deep experience and a narrow range of outcomes.

In many other areas, including stock markets and the state of the world at some point in the future we are prone to irrational thinking, owing to the large number of variables and our own biases. In decision-making, too, we are often led astray by our biases. For instance, we confuse the process of making a decision with its outcome. We don't recognize that the decision and its outcome could be disconnected.

When the outcome is seen as good, we may conclude that we made a good decision, but that needn't be the case. We may simply have been lucky, and our decision-making prowess may have had nothing to do with how the decision turned out. The only thing we can reliably do is to make sure that we have followed a rigorous and high-quality decision process— one that entails collecting a reasonable amount of data, involves the appropriate experts, considers a set of clear alternatives, and considers and attempts to address typical biases that we are all prone to.

These cognitive biases are mind bugs, which tend to make us believe that we are more skilled than we actually are. They lead us to judge ourselves more favourably; to rationalize and excuse our own actions more easily. Consider the well-documented Dunning-Kruger Effect, which is based on the observation that we consistently judge our own abilities and competence more highly that that of others, which can, in turn, blind us to our own shortcomings.

Being in a position of power also makes us less self-aware—possibly because we get less honest feedback from others to correct our own, biased self-perception. Further, having experienced an event makes us more confident in predicting the outcome of such events in the future. These are just some of the biases that distort our self-perception and lead us to act in ways that are, at best, counter productive and at worst harmful to ourselves and others.

As individuals, we strive to maintain a cohesive narrative about who we are. This narrative identity often leads to phenomenon such as avoiding cognitive dissonance and confirmation bias. This means, among other things, that we avoid anything that may suggest that we are wrong, whether about who we are, what we know, or what we have done. At the same time, we selectively focus on and incorporate information that supports what we already believe.

Our biases may be both innate and socially conditioned. They include our preference for family over others, as well as our tendency to form groups and to be suspicious of those outside our group. This may be extended through socialization, which, depending on the specific groups we associate with, may lead us to think of others as being less than us, even less than human. And none of us is immune to these bugs.

Our biases are often implicit. That is, we are not consciously aware of them. Yet they influence our perception and our actions; and they impact other people, as when we cross the street to avoid a presumed threat; treat someone as if they were unworthy of our attention or compassion; when we don't give them a job or hold them back from a promotion. Biases are the underlying mechanism of racism, sexism, and other forms of discrimination. Even people who are born blind are not necessarily blind to race— they are readily socialized to take on biases, just like someone that can see the colour of another person's skin.

Understanding ourselves requires understanding our biases— because we all have them— and countering them where it is beneficial and possible to do so, keeping in mind that it is not always easy to even recognize our own biases. And the way to start is by engaging others to understand ourselves— because others will be more able to see our errors than we are. Asking others for honest feedback on our behaviour is a powerful way to start seeing ourselves more objectively, to start seeing ourselves as others see us.

When we begin to comprehend what and how we are to other people, we can not only understand ourselves better but also begin to build more meaningful relationships. Indeed, to understand ourselves we need to see ourselves more clearly, without the rose-coloured glasses, without slipping into the habit of needing to feel better than others— to feed our ego to have self-worth. Often our anger, resentment, and ego blind us— we project those feelings onto others, attribute intent that may not be there and respond to it in our own behaviour. All of this affects how we relate to each other.

8

Navigating Problem Space

The modern economy, which in the 21st century is dominated by the so-called knowledge economy, places a great emphasis on intellectual capability. Many Western schools identify "gifted" students and stream them into separate classes with advanced curricula. Elite primary and secondary schools, as well as universities, similarly select students based on demonstrated academic ability.

The criteria often used for such selection include grades and IQ tests. Despite their controversial history, IQ tests in particular, have continued to be a key tool for those wishing to separate the brilliant from the average. The underlying assumption is that "general" intelligence is a predictor of success and can be measured with the right tests. But what was once a point-in-time assessment intended to identify children's developmental needs became part of the quantified life, a definition of one's worth as a human being— essential to how we define ourselves. It is a new caste system— one that isn't based on what work we do, or which group we were historically part of— but on a supposedly-objective measure of our intelligence.

The growing human tendency and ability to measure everything and assign it value, even those things that are not easily quantifiable, sets us up for a world that can be manipulated to support specific goals. It can be used to decide who goes to the best schools, gets the good jobs (or even the type of job they get), how resources are allocated and ultimately, how wealth is distributed.

It can also be used to enforce social structures, to keep people in their place, in as much as status has always done, through systems of class and caste, for instance. It can be used to justify discrimination and sometimes even worse— policies that decide who gets sterilized or who gets criminalized. It establishes a self-perpetuating system that soon becomes difficult to dismantle. It engenders inequality for generations.

And the natural course of such quantification leads to labeling people for life, making it difficult for them to move beyond whatever category they have been assigned— retard (now intellectually disabled), average, brilliant, and so on. This then gets tied to DNA and further categorization by race and ethnicity— it becomes destiny.

However, this does not mean that the study of intelligence and attempts to quantify the seemingly unquantifiable are necessarily wrong-headed. It just means that we must take great care to avoid over-generalization, simple explanations, stereotypes and cognitive biases.

It also important to not get trapped in a too-narrow framing of intelligence. To some extent many of these tests assume a shared cultural and linguistic context. It is not unlike the children's growth charts that pediatricians use. The percentiles, which parents often obsess over (my child is in the 5th or 90th percentile for height, weight, etc.) are based on a sample that is increasingly unrepresentative of the population— geographically, racially, genetically. To the point where it provides little in the way of practical information.

Psychometrics are perhaps about as dangerous when they claim to accurately measure aptitude— be it intelligence or sales skills. Our definitions of these things are often flimsy, having little or no scientific basis. But the tests become widespread, and they are treated as gospel because the alternatives are too complicated. They become axioms by which we judge someone's entire existence; impose life sentences based on presumed capabilities and limitations.

But context matters. Perhaps a better way of thinking about intelligence— if we even want to use the term— is to consider the idea of fitness based on an organism's environment. An organism— or a person, if we are talking about humans— can adapt, survive, and perhaps even thrive if its abilities are suited to the natural environment in which it finds itself. This is perhaps more straightforward for animals that don't voluntarily or frequently encounter new surroundings. Such animals aren't typically subject to drastic changes in the environments they do inhabit— barring the occasional natural disaster or human intervention.

Humans, however, and modern humans in particular, can readily change their environments and can easily move between environments. This means that what works in one context may not work in another. This is where the developmental psychologist, Howard Gardener's theory of multiple intelligences presents a useful contrast. By his account, intelligence is not some monolithic quality that we either have or don't; it is not something that can be measured by a simple standardized test.

Instead, Gardener posited that intelligence comes in many flavours, or "modalities." He suggested that musical, athletic, inter-personal, visual or other abilities are also forms of intelligence. He sought to capture a more comprehensive range of capabilities that could be used to assess someone's suitability to a certain environment. For instance, being physically intelligent may make us suitable for a career in athletics or home building or other areas where physical prowess offers a clear advantage.

Similarly, having inter-personal intelligence may help us form relationships and work with people more easily. He identified eight specific types of intelligence, including— in addition to the ones already cited— linguistic, mathematical-logical, intrapersonal, intellectual, and naturalistic, though there could easily be more. Critics pointed out that these were simply talents or personality traits. But couldn't that be said about general intelligence as well? If Gardener's definition was too broad, then the traditional definition is too narrow.

Regardless, his point wasn't that a person could only have one type of intelligence but that we may all have bits of all of them— just to varying degrees. If nothing else, Gardener's definition offers a more equitable way to assess humans precisely because it is more expansive and thus accounts for the diversity of human abilities and environments.

Today psychologists talk about intelligence as the ability to navigate problem space— that is, as someone's depth of resources and degree of creativity in solving a particular kind of problem. And problems don't just mean intellectual puzzles, but spatial, musical and other kinds. Regardless of the chosen definition, what does it mean to put a certain emphasis on intelligence vs. other aspects of humanity, including how we behave towards each other. What role should our intelligence— generally or multiply defined— play in how we see ourselves? In how we compare ourselves to others? In how we judge humans?

Measures of intelligence may have some practical value, but do they have moral value? Does having greater intelligence make us better or worse humans? These are the types of questions we need to answer for ourselves as we set out to understand who we think we are and who we would like to be. Intelligence of any kind may be determined both by genetics and the environment— by nature and nurture. And aspects of both may be out of our control. If so, we must decide how much that means to us— how much or how little it defines each of us as individuals.

9

Feeling Like Some Body

We have gone from dismissing emotions as the provenance of women and irrational people to thinking of them as an integral part of who we are. We believe that we have emotions, that they are important, and that they reside in the "feeling" parts of our brain, separate from the rational, "thinking" parts of the brain. Chief among these is the amygdala, which appears to play a primary role in habitual behaviour. It is the part of the brain that uses sensory inputs to continually scans the environment, remains vigilant, and prepares to respond to any perceived threats.

The author, Chip Conley, has suggested that emotions can be broken down into their constituents to better understand their origins— like an equation where 5 is a combination of 2 and 3. So, for instance, despair = suffering - meaning. While this is helpful to a degree, it is also incomplete. Just as in mathematics, equations can sometimes have more than one solution. Further, emotions—unlike numbers—can have multiple fathers, mothers, uncles, aunts. Emotions are the product of complicated processes within the body and the environment within which it finds itself.

The neuroscientist Lisa Feldman Barrett has posited that emotions don't even exist. Instead, she suggests, they are the brain's interpretation of different body states, as it tries to achieve allostasis. This builds on neuroscientist Antonio Damasio's theory that emotions are the body's response to various stimuli, while feelings are the conscious awareness of these responses.

If scientists like Damasio and Feldman Barrett are correct, then perhaps there is little value in therapy or reflection to better understand our emotions. We cannot plumb the hidden depths of our minds to figure out why we feel or act a certain way because those reasons don't exist, or at least, they are much simpler and mechanistic. This idea builds on 19th century theories developed independently by William James and James Lange and now known as the James-Lange Theory of Emotion, which holds that emotions are responses to body states. In William James's view an emotion is only an emotion when we feel it—that is, when we have a conscious experience of it, which agrees with Damasio's theory of consciousness.

At one level, what we experience as emotion is not so easily described in terms of brain processes or even identifiable as distinct phenomenon. Although functional MRIs can be helpful, they are also reductive—characterizing all brain activity in terms of blood flow and oxygenation. That is, if blood flow increases to a certain part of the brain when someone claims to be experiencing a particular emotion, then that part of the brain is said to be "activated" and is assumed to be responsible for originating and managing that particular emotion. However, MRIs have limited resolution—blood flow is not the whole story—and we have yet to find a way to monitor cellular-level changes while someone is being subject to an experiment.

Again, there may be truth in these MRI findings, but the technology also presents a rather simplified view of brain function. For instance, beyond brain activation, emotions also exist in the context of language— that is, in our ability to describe them. This is where, we see disagreement about exactly how many emotions there are and where their boundaries lie. For instance, Germans describe schadenfreude— the pleasure that comes from someone else's misfortune— as an emotion that has no equivalent in English.

The American anthropologist, Rentao Rosaldo describes another emotion— *liget*—expressed by the Ilongot tribe in the Philippines. At first, he was unable to completely comprehend it except to convey the tribespeople's desire to "take a head" when they were in the throes of this emotion. Later, he came to experience it directly after his wife died in an unfortunate accident on one of their research trips, leaving him with two young children and an ineffable sense of loss. When, one day, he found release in irrepressible howls of grief, he describes it as a high voltage current running through his body, triggering both anger and grief.

Our inability to precisely define emotion aside, there is also a difference between the physiological responses triggered by an emotion like fear (sweaty palms, racing heart) and the conscious experience of the same emotion, which may happen even in the absence of any actual threat, but which cannot be readily measured.

But is this so? It is easy enough to identify instances where we discover ourselves gritting our teeth before we realize we are angry, or of having a queasy feeling before we are aware of fear. From an evolutionary standpoint this makes sense, as even animals that we don't think of as having self-awareness appear to exhibit emotions such as fear. Thus, we could have emotions without even being aware of them. This is evident in the way our moods— often unconsciously— predispose us to reacting in certain predictable ways to events. For instance, we may find something pleasant if we are in a good mood but less so if we are not.

Additionally, we tend to have a more negative emotional response when we are feeling stressed or threatened in some way. We are more easily triggered— our thresholds become lower, and we may lash out and overreact. These are the times when we could benefit from self-reflection but also the moments when it is hard to practice it.

Stress and other emotions can both trigger and be influenced by hormones, including ones like cortisol, oxytocin and estrogen, which can also mediate the production of neurotransmitters like serotonin, dopamine, norepinephrine and dopamine.

Interestingly and perhaps unhelpfully, our own heightened emotional reactivity triggers a similar response in others. Even at the biochemical and neurological levels, we can initiate similar processes in others through aggression, negativity or even excitement. Consider how excitement or panic in a group or crowd can be "contagious." This is where System 1 takes over and, if not checked, can seriously damage our relationships with others— a form of mutually assured destruction.

Regardless, the simplification of emotions to equations does provide us with levers we may be able to manipulate to change something— if that is our desire. At a very basic level, what we feel about someone, or something is regulated by what we find pleasurable and what we expected compared to what is or has happened. For instance, we may be happy when things go as planned, or when something unexpected happens that is pleasurable. Similarly, anger and disappointment may be the result of missed expectations and understanding this may help us to better prepare for regulate our emotions.

Ultimately, emotions appear to be an integrated conscious experience of our body's state— internal and external— at a given moment, interpreted through the lens of our perception and thoughts about where we are, whom we are with, what's going on at that moment and what we are thinking about. All of this could also be recursive, meaning that as new information flows into the brain and it updates its predictions about the state of the body, the emotion it experiences also changes.

10

Why Behave

Among all the things that shape our identity, how we behave is key to how others see us and even to how we see ourselves. Our behaviours are the outputs of neural processes. They are determined, in part, by our brain's model of its world, built atop memories, and by its attempt to correct model errors using sensory data, as well as by both conscious and unconscious processing of these inputs.

Genes underlie all of these, determining the quality of the inputs and quite possibly how they are weighted and processed. At the neural level, our behaviour is the result of interactions among the amygdala, insula (sensory binding, motor control, emotions), hippocampus (memory) and parts of the frontal cortex, which attempt to coordinate our responses.

So how well do we behave? On one level, this question is meaningless without reference to socially accepted norms. On another, we behave in a programmed manner, responding to stimuli without giving it much thought— as when we duck for cover upon seeing a dangerous object hurtling towards us.

How we act in the world appears to be highly dependent on the context. All of us behave badly—as measured against prevailing social norms—at least some of the time. This may have more to do with power, opportunity, and threat perception than with any inherent moral failing. The rationalist view is that morality is not an absolute and permanent quality that some posses in greater quantities than others. It is, rather, situational, and therefore fungible. And one situation that consistently seems to encourage bad behaviour is power.

However, if power corrupts, does it also corrupt our morals—our propensity to treat other people badly simply because we can, because we are human? Perhaps so, and possibly for no other reason than the fact that we are dopamine addicts. We seek the pleasure that results from dopamine release in the brain while avoiding pain. We are also hardwired to pay attention to status and hierarchy and spend significant parts of our day engaged in gossip— often negative, and almost always about other people. It reinforces the "othering" that we engage in almost instinctively, organizing peoples into those that are with us, and those that aren't— our in-groups and out-groups.

Yet, we also have a capacity to overcome our built-in biases, our acculturation, our genetic propensities. Even genes do not seal our fate or guarantee our destiny. Experience, we are increasingly learning, can affect the expression of genes, and thereby how we behave in the world. Our behaviour is the result of processes that we are not aware of. We could think of these processes separately as thinking, feeling, perceiving, etc.

But these could all be one process— the process of thought itself. Or more accurately, behaviour is nothing more than the brain's response to its model of the world, which is updated throughout our lives. Behaviour is action that is intended to keep us functioning— keep us alive long enough to pass on our genes and take care of our young. It is the brain's attempt to achieve "allostasis" (a state of optimal operation) by adjusting whatever physiological parameters are necessary to do so.

It is perhaps in our nature to believe that we behave well, that we are right most of the time. When we misbehave, we tend to rationalize, to dissemble— not least to ourselves. But behaviour— ours and that of others'— is complicated. It is the inevitable product of our history, including memory, perception, genes, and evolution. But we know that memory is flawed, and perception is likely only a representation of the world. Thus, we base our responses— our actions— on what we perceive and what we remember, neither of which can ever be objective. And often, what we remember is what we have learned— essentially a pattern of neuronal activation that is intended to repeat a particular behaviour in response to a particular stimulus.

Learning itself appears to occur when the neurotransmitter, dopamine, is released in response to a "reward" that occurs, often unexpectedly, signaling to the organism that the behaviour may be worth repeating. The theory behind this type of reinforcement learning (which is also used to train artificial intelligence algorithms) is called the reward prediction error coding hypothesis. This builds on work in classical conditioning by the 19th century Russian physiologist, Ivan Pavlov, who demonstrated that dogs could be trained to salivate in response to a bell if a bell was struck each time, they were fed. Thereafter, the dogs would salivate even if the bell was rung without it being followed by food.

Beyond such experiments, some of our best understanding of ourselves comes from instances where we do not behave as expected— that is, when our brains and bodies misbehave. For instance, we are constantly categorizing and judging other people, by how they look, what they wear, how they speak and in many other ways.

The hormone oxytocin influences social cognition— that is our sense of who is within and outside of our social group. Similarly, studies with adrenaline suggest that people's interpretations of their emotional states can be largely influenced by those around them. The same shot of adrenaline can make a person feel fearful or angry depending on how other people around them are behaving.

Most intriguing is the role of the neurotransmitter, dopamine, which appears to be important for behavioural control and yet has different effects on different people. It may also play a role in determining which thoughts become conscious and are acted on and which are ignored. It may also control the motivation to act at all.

In general, we tend to be more prejudiced against people in the out-group than in the in-group. This "pair-bonding" with members of the in-group has its uses— for example, mother and child; family, tribe, etc.— as group members are predisposed to help and protect each other from external threats.

Yet we judge each other as if we ourselves are infallible, morally incorruptible, mostly right. This sets the stage for uncivil discourse sparked by moral indignation, which prevents any real arguments from being heard or debate from happening. There seems to be an inexorable rightness to our own thinking and actions that isn't there when we consider other people's.

When we say or do something, it feels entirely reasonable. It is almost impossible not to feel this way, to not judge ourselves more favourably, to not give ourselves the greater benefit of the doubt. Granted, depending on personality and situation, all of us may judge ourselves more harshly from time to time, or even a lot of the time.

Behaviour is also inextricably linked to the notion of free will. That is, do we choose to behave in a certain way or are we compelled to do so by our experience, environment, and genetic inheritance? It is clear and inarguable that our genes are important—DNA is ultimately transcribed into proteins, which in turn determine cell function. The accompanying biochemistry does, to a large extent, control how we behave. And as far as we know, we have practically no conscious control of these processes.

At the same time, gene expression— that is, which proteins are ultimately produced— is also determined by our current environment, as well as by the environment of our ancestors, which determined which genes they passed on to us. The impacts of food shortage, conditions in the natural environment (amount of sunlight, availability of warmth) as well as the presence of threats, all determine which genes are expressed in our own lifetimes, and which, in turn are transmitted to our offspring.

Further, other people's actions (including parents, friends, society-at-large) can change our environment and thereby influence our response to any particular stimulus. Given all this, can we truly have free will? It is an open question that continues to be debated by philosophers and scientists and a question to which we may never have a clear answer (for an in-depth discussion of free will and its implications, read Determined by biologist Robert Sapolsky).

Nevertheless, each of us has a sense that from moment to moment we have some control over our actions; some ability to choose—and perhaps we do within bounds. After all, if we consider the probabilistic nature of the most elementary forms of matter (quarks, electrons, etc.), it is conceivable that this property would be present at the macro level of cells and organisms. However, to-date there is no evidence that quantum phenomenon (such as superposition, discussed later) are present at larger scales. Presently, quantum effects have been demonstrated in crystals with up to 10^{16} atoms.

Speculatively, however, our behaviour could also be probabilistic in as much as the exact position of an electron is. That is, we may have a range of potential reactions to any given stimulus and how we ultimately act is determined by the context, which leads to a "collapse of the wave function" (something we will explore later), which leads to the manifestation of one outcome from a range of possible ones— and one behaviour rather than another.

11

Who and Whom

If biology is the hardware of our being, then identity is its user interface. It is the set of assumptions that informs our interactions with other people and the natural world. Our identities are complicated constructs— biochemistry shaped by ecology, including human sociology.

William James famously distinguished between the "I" and the "me," the "knower" and the "known," the "thinker" and the "thought about." For James, the "I" was related to subjective conscious experience while the "me," was related to how we think about ourselves in the third person when recalling memories in reflecting on past as opposed to current experiences. By that definition, identity spans both the I and the me.

Each of us takes a different aspect of ourselves— say our sex, gender, nationality, race / ethnicity / culture, or profession and uses it as a scaffold on which to build an understanding of ourselves. But while these are important, they create the flimsiest of foundations. They lead us to think that we know a lot more about ourselves than we really do.

Take nationality for instance. This is often nothing more than an accident of birth, yet it is something that defines whole parts of our existence, including what kinds of opportunities we have access to; the degree to which our ambitions can be thwarted by forces beyond our control; and our life expectancy.

We have now become accustomed to the idea of identity as the thing that is ultimately at the core of our being. However, when someone says this is what it means to be us— whether that's a brown man in Britain; a gay Chinese man in St. Petersburg, a Ghanaian woman in Singapore— we should be skeptical. They are conflating reality with identity. Perhaps it is *a* reality but not *the* reality, which is much more complex and irreducible to sociological categories.

We take these incidental labels and generate boundaries that we use to decide who is with us and who is not— who is a friend and who is foe; who is deserving of our sympathy and solidarity and who is not; who is an ally and who is an enemy; whose consciousness has been raised and whose is irredeemable.

This is not new, we seem to have an evolutionary preference for our own groups—although humans, unlike many animals have a multitude of groups that they can be part of through the course of their lives. But even if such automatic, implicit, and often subconscious categorization is rooted in biology, it seems to be something we can overcome— perhaps by changing the labels, by altering the boundaries of us vs. them.

To state a tautology, social identity is, obviously, socially constructed—it does not exist in any objective way outside of a social or cultural context. To varying extents, it may even be imposed by others, although no individual or group can claim to be the imposer or the imposed. It is an ongoing, dynamic, and evolving exchange that defines and re-defines each of us and the way we engage with our surroundings.

And although it may be inorganic, it is no less real. In one sense identity emerges from biology in the context of an organism's ecosystem. Statisticians and computer scientists speak of a Markov blanket (named after the 19th century Russian mathematician Andriy Markov who studied random processes) as a separation between states or nodes. In biology, Markov blankets define the boundaries that circumscribe an organism in an environment. As humans, we have Markov blankets around individual cells but also between our bodies and the outside world. These boundaries are statistical separations that are independent of each other.

Perhaps we can use this idea to define ourselves apart from other people, as well. After all, Markov blankets separate internal states from external ones. Similarly, our identities are necessary to our psychological integrity. Indeed, as noted previously, one of the postulated mechanisms of schizophrenia has to do with the brain's compromised ability to separate ourselves from others— our identify from theirs. As with cognition, if there is one area of the brain that is crucial to our sense of this separation, it is where the temporal and parietal lobes meet— the temporoparietal junction (TPJ). The TPJ has been implicated in disorders such as schizophrenia and Alzheimer's. And it is through these diseases that we have insights into its role in our sense of self.

For instance, damage to this part of the brain and another part called the insula— as well as electromagnetic stimulation — has been shown to affect moral judgments and the ability to distinguish the self from others. Manipulating it can also trigger out-of-body experiences and a feeling of presence (i.e., hallucinations of ghosts). Similarly, mountaineers sometimes experience a presence (a "third man") under extreme fatigue at high altitudes, where both the landscape and lack of oxygen contribute to the visions.

The TPJ is a clearinghouse, where information from both outside and inside the body, including other parts of the brain are collected and processed. It is, in essence, the interface of consciousness— where our minds meet the physical world. And when it is functioning within normal parameters, we feel ourselves as having a defined self.

That self is often based on a well-rehearsed origin story and a personal history that is continually being written and re-written. Different aspects of the story come into play each time we engage with something outside our minds— particularly, our fellow humans. Thes stories help us to not only relate to other people but to ourselves as well.

Thus, our stance towards others is determined by both our biology and the environment in which we grew up— not just by genes. If anything, while genes may predispose us towards one tendency or another, the degree to which a behaviour is expressed could (depending on the specific characteristic) ultimately be influenced by external factors.

So, psychology and sociology are intertwined. But the range of variables is vast and despite the steady stream of psychological research findings, the best evidence for why we behave as we behave seems to be correlational and likely subject to statistical biases. Regardless, such studies point to the variety of identities we may adopt given a set of starting conditions (biological and sociological).

In the twenty-first century, issues of identity have gained prominence through the growing acceptance of groups that were previously marginalized because of their sex, gender identification, sexual orientation, or other characteristics. What was once thought to be lifestyle preferences became acknowledged as something more intrinsic, perhaps even genetically determined.

What remains debatable is whether such self-identification can cross other seemingly harder lines, such as race— that is whether, for instance, a white person can identify as black or vice-a-versa without a genetic basis. For now, the answer seems to be no—so much so that anyone that tries to cross those boundaries is often deemed to be engaging in cultural appropriation. This suggests that the lines between biology and sociology haven't been clearly defined— some aspects of identity are subject to choice, while others are seen as organic and immutable.

Beyond these areas, our identities extend to encompass other parts of our lives, including associations, both relational and physical. That is, both people and things. We derive part of our sense of self from the people and communities we live among or choose to interact with. It could go both ways— we may seek out specific individuals or groups because we believe ourselves to be certain kinds of people. Conversely, we may come to think of ourselves differently based on whom we mix with.

It is curious that the present use of the word identity and its prominence are a post-World War II artifact. Before then, our immediate neighbours and environment primarily determined how we thought of ourselves. Thus, loyalty to family and tribe likely outweighed any loyalty to country, religion, or race. Granted that many of these groups were so homogeneous that the likelihood of a family or tribe having members of difference races, religions or ethnicities was probably small.

These tribal affiliations can confer status and boost the ego, or they can induce feelings of shame and inferiority. Some of us may be able to shed or form such associations more easily than others. But regardless of their origin or nature, our affiliations influence our behaviour. Consider, for instance, that teenagers are more likely to engage in risky behaviour when they are with their friends; or that some people may feel the need to acquire material goods to emulate the class of people they aspire to join.

The wider culture, media representations and social norms also influence identity and the degree to which we are comfortable with it. Generally, if we feel like we fit into a specific group, we will be happier than if we feel like impostors who do not belong.

Beyond the identities that we form to define ourselves within our communities, we also rely on physical objects and notions, such as institutions and achievements. For instance, the various objects we own, the clothes we wear, the way we style our hair, treat our skin and so on all play a role in defining us for ourselves and others.

Similarly, there is a part of our ego that exists only by association— with the institutions that we have been a part of or the places we have worked; by the organizations that we belong to, the protests we join or the causes we support. We could say that we each wrap ourselves in many, layered Markov blankets, all of which bound different aspects of ourselves. Our identities, then, are a set of nesting dolls, each presenting a different face to its selected audience.

Individual experience, aside, the collective experience of the society in which we live— its culture—also has profound effects on our concept of self. Even our individualistic vs. collectivist tendencies may be culturally determined. How individualistic we are— and perhaps even some of our political preferences —may influence how we interpret memory, events and outcomes.

For instance, psychologists such as Michael Varnum, Thomas Talhelm, and others suggest our social orientation (how we see ourselves and others) is influenced by our natural environment and influences the way we think. For instance, being explorers—frontiers people—may tend to make us more individualistic than if we were farmers dependent on each other to harvest rice. As farmers, we would ensure that our own irrigation efforts did not damage our neighbour's field. Similarly, if we were descended from nomads, we may be more likely to emphasize the importance of enforcing honor (including through honor killings) than if we were farmers, whose crops are not as easily stolen. This is possibly because each orientation is beneficial to its specific context.

Collectivism makes us see the world as more interdependent, potentially predisposing us to cooperation. We are then less likely to see our own roles as significant and consequently would be less negative in our interpretations of events. This may also lead to an underestimation of our own abilities but also confer protection from depression. In traditionally collectivist societies that sense of interdependence and responsibility extends beyond the living, to incorporate ancestors and future generations. Indeed, collectivism could predispose us to see things in context, to look for relationships among things, rather than focusing on individuals or individual elements. At the same time, it may lead to the suppression of individual needs and desires to serve the greater good.

By contrast, individualism is useful in exploration, entrepreneurship—activities that entail risk-taking. But it also makes us prone to ego-inflation and over confidence—to assume that our role in events was more important that it may actually have been. We will tend to underestimate the role of luck or the contribution of others, especially when it comes to whatever successes we have achieved.

Whether individualist or collectivist, we often end-up defining our self-worth according to our culture and specifically by what it holds dear. We use its yardsticks—explicit or implicit—to measure ourselves. And sometimes we may feel like outsiders who don't fit into our own culture—at the broad social level or the micro level of our day-to-day professional or personal lives. We may even have a distinct sense of being an imposter (imposter syndrome), only pretending to possess qualities and abilities that fit the cultural norm.

Based on our heritage, our habitat and our experiences, we tell ourselves stories about who we are. I am this type of person—a visual learner, say, or a people person. We may say to ourselves, I like cricket but not tennis, steaks but nor burgers; I am smart, or I am dumb. But what if we told ourselves a different story. Would that change anything?

The words we use to imagine ourselves have power and there is some evidence that using different words, or at least not being defined by restrictive words sets us free to explore different ways of being. This is not to say that we can be anything we want to be—that our abilities have no limits. Clearly, we do have limits—both physical and psychological—but many are not as inflexible as we may think. Physical exercise can make us stronger; mental exercise—for instance in the form of meditation—can make us calmer; cognitive behavioural therapy or self-talk can improve our mood.

The reality is that our self-conception and other people's perceptions won't always agree; and if they do it's not necessarily because both sides recognize some ultimate, incontrovertible truth. It may be because they are predisposed to agree with us. That is, their self-perception benefits from agreeing with us about ours. Each one of us does many things to maintain our individual self-image, our notion of the kind of person we are. For instance, we may be more inclined to do some things because it would bolster that image— for ourselves and for any observers.

We may be motivated, in large part, by a desire to be liked, a need for validation. So much of our ego, our sense of self worth is determined by external factors, by what other people think of us, or what we believe they think of us. We are all susceptible to this. Over time, the need for approval may diminish as we mature and grow comfortable with our own nature and situation, but other people's views are likely to continue playing an outsized role in our satisfaction with our life and our ego. What academic degrees we have earned, which schools we attended, which organizations we worked for, may all remain important, if to varying levels and in different contexts.

And we judge each other, mercilessly, unwittingly; even daily. We do so in an attempt to better define ourselves by differentiating us from them. We judge and are judged based on looks, colour, height, accent. It feels like an inescapable part of human existence. Despite our best intentions, social evolution has not trumped biology. not yet. Clearly, identity plays an important role in our lives, in our self-conception, and it almost seems to have an inevitable hold over us.

Despite its many limitations, psychology is a useful lens to peer through when we are trying to understand who we are and why we behave the way we do. Especially since the 1970s, the field has made tremendous strides, pointing out the irrationality and contradictory nature of our behaviour. It has revealed the flaws underlying economics, specifically the assumption that people behave in rational ways to maximize their individual outcomes or utility.

As we have discussed earlier, one especially interesting idea to come out of behavioural economics is the notion of two separate thinking processes, System 1 and System 2. Some of this originated with the dual process theory advanced by psychologists Peter Watson and Jonathan Evans in the 1970s. But it goes back even further to William James, who many consider the father of psychology.

Regardless of their specific formulations, what these theories have in common is the idea that our thinking falls into two distinct types: automatic (and often irrational) and considered (and typically slower, more rational). This implies that our brains are often lazy and rely on heuristics— shortcuts to knowledge— to form judgments. Thus, our first impressions are significantly biased by habits of thinking we have accumulated over a lifetime. In fact, we are loath to think critically, if we can avoid it. But critical thinking, like other habits can be cultivated.

Nonetheless, critical thinking is not a cure-all— we will still fall prey to shortcuts. We will be too easily influenced by cues— we will click on a link we shouldn't have clicked on. We will believe something we don't have enough reason to. We will trust people we don't have good reasons to trust. This is partly because, as the 18th century philosopher, David Hume, put it "reason is and ought only to be the slave of the passions."

Yet, we will not always see ourselves as being subject to such perceived flaws. This is partly because of the narrative fallacies we are subject to— the stories that help us interpret the past and set our expectations for the future. In many ways, we experience the future we expect to live in, based on the stories we have told ourselves.

These characterizations sometimes take the form of categorization. We split people into good and bad, honest, and dishonest, criminal, and innocent. More often than not, we put ourselves in the good, honest, innocent categories. Even when we make errors that may make us seem the opposite. We excuse ourselves; we rationalize.

But others must remain in their categories, unless we get to know them at a deeper level, which rarely happens with people we hear about but have no prospect of getting to know in person. So, our tendency is to lionize and excuse those we admire and demonize and condemn those we do not. There is little room for nuance. What we miss in all this is that all of us may at different times, and in specific circumstances, exhibit behaviours that others may consider "bad," for cultural, moral or other reasons.

Indeed, psychologists suggest that most behaviour is situational. And comments like "that seems out of character" for so-and-so are often made because many of us live in stable environments. Thus, we mostly tend to see people in predictable situations where they seem to consistently exhibit one kind of behaviour and a fixed set of values. When they encounter a new situation, they may behave in completely different ways that would be considered "out of character."

Psychologists say that personality predicts the statistical average behaviour over a lifetime, whereas the situation determines the behaviour we can expect to see in a given moment. This is why it makes little sense to describe people as being one thing or another; or to permanently relegate them to categories that could encourage othering and discrimination.

Nevertheless, and however flawed, we each have a strong sense of who we are. And we become uncomfortable when that self-conceptualization is challenged by other people or situations. While on the one hand the brain appears to rely on error correction to continually refine its model of the world, it seems to balk at perceived over-corrections of the model's perception of itself.

Why do we bristle when people suggest that we are other than we seem, than we strive to seem? Perhaps, we take offense because it unsettles us. It is akin to shaking a fishbowl, which dislodges the sand and small pebbles, the sedimentary layers of our self-image that have been laid down over many years. It muddies the waters that we try so hard to keep clear, prevents us from seeing ourselves as we want to. We don't want to let go of this singular, unwavering image— this identity we have created for ourselves.

Perhaps it is also the feint towards a truth we would rather not acknowledge, even to ourselves. The truth that we are not wholly who we appear to be. The truth that we are performing a version of ourselves. The truth that there is a disconnect between what we think and feel in a given situation and how we actually speak and act. It is hard to live with that cognitive dissonance, so most of the time we pretend that it is not there.

When someone challenges us, when they point out what's hidden under the skin, we can't pretend any more. And often our response is to lash out in defense because we do not want to let go, do not want to admit— or even acknowledge to ourselves— the existence of another version. We'd rather fight even if we know it's not a fight based on merits, maybe not even a battle worth fighting—we fight, nonetheless, to maintain the integrity of our self-conception.

At the end of the day, it is perhaps useful to remember that the map of our identities is not the territory of our selves— if only because there is no singular self.

12

Where We End

In his book, Being and Time, the German Philosopher Martin Heidegger, suggests that being is time and time (at least for humans) is a finite thing. Thus, it is only by admitting our mortality that we can truly start to live, be authentic. Imagine a roller coaster ride, a bungee jump, sky diving— these experiences bring us close to death, yet make us feel alive, energized. Being in the face of death, going to the edge—this is what enlivens us.

Life as we know it presupposes death. To be alive is to accept the risk that we will not be alive forever. Thus, life itself is the greatest risk we could possibly take, and we are already taking it— simply by living.

Further, regardless of personal beliefs about whether death is the final act, we can at least agree that this life will end in death. It is powerful knowledge. And if we can remain cognizant of this fact, we can find a unique kind of motivation for seeing ourselves as we are and perhaps even for making life more meaningful.

But what does it mean to find meaning, to find our true selves? Is there truly a constant, authentic "me" that each of us can even discover? What does authenticity mean when our personality may not be fixed, when our values and our behaviour may be situational, inconsistent? Does it mean that we can never be authentic, that we can never have a fixed, coherent identity?

Besides, is there only one, truly authentic version of each of us? Or does it only seem so because we mostly live in stable environments that allow us to behave consistently?

Do we not, after all, play different roles in different settings? Aren't we different people at different times and with different people?

If we are simply playing a role, which entails a series of actions set in motion by a prime move, billions of years ago then is there even a point in identifying ourselves as entities, separate not just from other people, but from other things, as well?

Further, are we just projecting authenticity, especially when many of us live public lives on internet based social networks? Are we just presenting an idealized version of ourselves, which while being partly true, rarely exposes anything raw or vulnerable—anything that opens us up to failure or rejection?

Any exhortation to be our true selves is misleading at best. At worst it sets us up in opposition to others— we may externalize our difficulties, blaming others for preventing us from being true to ourselves. Yet, if the self is meaningless without others, without the world, then it is the self in context that is what makes us who we are.

Similarly, the encouragement to follow our passions assumes that without them we will be unhappy, untrue to ourselves. This presupposes that our interests will remain stable. It doesn't consider that what we value today may not necessarily be what we value tomorrow. Our "passions" or more generally, our interests, may, and likely will, change over time. The world will change, and we will change along with it.

We need to recognize that the work of living, the messy business of relating to others and to our natural environment cannot be abandoned in pursuit of some personal nirvana. Being human comes with responsibilities to each other and while we can expect and encourage others to live up to their responsibilities, we may not be able to demand it. So, we have to make a choice about whether we should do our own duty without knowing whether other people will do theirs.

And death is the ultimate equalizer. It represents the termination of all responsibilities. It resets everyone's personal credits and debits to the point where they are meaningless.

Despite this, we venerate the dead in a way that we don't always laud the ordinary living. We call the dead beautiful people; we say they had a good heart; the loveliest; a kind soul. Is it only in death that people become saintly? Is it not disingenuous to offer such insincere praise?

Should we not do more to recognize the living but even then, see them as simply human, faults and all, and not as either extraordinary or subpar— just human.

And if most of us, on balance, are unremarkable shouldn't the dead, too, remain so? Why succumb to the tendency to make someone seem more than they were simply because they have died? Why not simply describe them as they were— remember the fullness of their humanity and their lives?

Each of us, too, will one day no longer be alive. If we wanted to be kind, we would say that the dead live on in memory. But with time, even those memories will fade. It will be the rare person that is remembered by history long after they draw their last breath. No matter who we are in life, no matter how successful, famous, wealthy, healthy or otherwise, this is our fate— to be forgotten.

It may seem futile to think about dying rather than living; but acknowledging our individual mortality can also be invigorating in unexpected ways. Indeed, the French Philosopher, Anne Dufourmantelle, advocated for taking big risks in life to make it more meaningful. After all, we have taken the greatest risk of all by being born and remaining alive. How can any other risk compare?

She can be lauded for taking that philosophy to heart, for in 2017 Dufourmantelle drowned while saving two children who had been swept into the sea by strong winds. However, should we, like her, risk life to live it? Should we not live life to the fullest before death comes for each of us?

But what exactly is a full life? A *bios telios*, as the Greek philosopher, Aristotle put it— may mean a long one (in years) and/or a complete one (in the variety and depth of its content). Length is easy to measure but content is harder. What counts as good content— the number and type of relationships one has? The accolades one accumulates?

The Roman Stoic philosopher, Seneca the Younger, thought that the problem wasn't the length of one's life but that so much of it was wasted, specifically on things that are meaningless (there's content, again).

It is curious then that Seneca's own life ended when the Emperor Nero ordered him to commit suicide after Seneca was implicated (wrongly) in a plot to kill Nero. But perhaps, like Dufourmantelle, Seneca died as he lived, in the pursuit of meaning.

Still, all lives, no matter how long or short; filled with meaning or wasted on inconsequential pursuits, come to an end. Does that mean that they are ultimately meaningless? In the grand scheme of things, in the multi-billion history of a universe that may well be infinite, will it matter that any one of us was alive?

The answer could well be that it does not matter.

And if that's true, the challenge for each of us is to continue living, to continue making meaning, without yielding to the temptations of nihilism, solipsism, or hedonism.

This notion is contained in the concept of dharma ("duty"), described in the ancient epic poems known as the Vedas. it is the idea that we all have obligations in life, to ourselves and to others. And we must fulfill those obligations regardless of how long we are alive. For even a short life can be a complete one. It, too, can be full, even if all one does is perform one's duties. That may be meaning enough.

Or not.

Some of us will want to wrest more out of life, we may want to do more than just our duty; or not do our duty at all— dharma be damned! Whatever we do— whether we pursue meaning or shun it; wish for a long life or a short, full one, we would be well advised to keep death by our side.

Doing so will remind us to measure our lives not against human yardsticks, but against the vastness and mystery of existence.

II

Belonging

13

The Centre of the Universe

Our default perspective is a self-centred one. We wake up and think about how we feel, we consider our thoughts, about what it is like to be us. It is all first person, subjective, solipsistic.

This is partly out of necessity. As autonomous organisms in the world, our aim is to survive and persist through reproduction. We desire to transmit at least our genetic information (but now also our cultural inheritance) towards eternity, with our progeny and their progeny as the messengers.

Why we do this is anyone's guess. We feel compelled to do as we do. Thus, we shouldn't self-flagellate about our focus on the ego. All the same, we easily become desensitized to the world around us, as we go about the business of living.

It is, as the author David Foster Wallace so memorably put it, like fish not being aware that they are swimming in water. In many ways we are not fully aware of our own environment. We are oblivious to the wonder of the world around us and the vast, unfathomable universe within which we exist. Even at a local level, at a smaller scale, how often do we look up at the sky above us, or the ground beneath our feet and wonder? How often do we ask the most basic of all questions. What is this? Why is this?

Perhaps this is an existential dilemma unique to our species. We feel so confident in the reality of our being, we cannot easily dispel the notion that our subjective experience as delivered through perceptual signals and interpreted by our brains reflects some ultimate truth.

We interact with other people and other living things. We interact with both animate and inanimate things— including the sun, moon and stars around us. But we cannot, at one level, be sure of anything except our own existence. The rest may as well be a mirage. Yet, we insist, cogito, ergo sum. I think, I feel, I experience, therefore I am. But are you? Am I? Are we really here? Is any of this real?

Even though this is an age-old metaphysical conundrum, it is useful to proceed under the assumption that, real or not, it helps to understand our environment. At the very least it is what we seem to depend on and therefore what we need to deal with, adapt and relate to, to continue existing, whatever reality is and however it comes about.

It helps to understand that we are in water and that the water forms an ocean. And that the ocean hugs a rock; and the rock is in space, orbiting a star; that the star itself is moving through a galaxy and that galaxy exists in spacetime, which is expanding and may well be infinite. That space time itself may be composed of quantum fields that sometimes manifest as particles of matter and sometimes as energy.

Even if discombobulating, we need to acknowledge these things. This is, acknowledge what we know, to the best of our ability to know anything at all. From there, it takes effort to move beyond ego, to correct the self-serving bias that prevents us from embracing what's outside of us, whether it is the natural world or other people.

It is harder when cultural and societal forces and trends push towards considering ourselves first— looking after number one, as the saying goes. Harder still when we are told to not just think about ourselves but to promote our brand, differentiate. This is supposedly the path to self-actualization, to realizing our dreams, to living our passions, to finding our authentic self— whatever it may be.

That, of course, is a recipe for narcissism, for willful blindness about what is going on around us. It is what would lead us to believe that we are not in water and perhaps even make us think that we are unique as a species and as individuals, existing in some special medium that is itself enhanced by our existence.

The antidotes to solipsism are various but at a minimum involve becoming more attuned to our surroundings, cultivating a higher degree of awareness— within and without. It entails listening to our bodies and thoughts and then spreading those tendrils of awareness outwards, pausing to consider everything we encounter— the sights, the sounds, smells, textures of things, the forces that push back against us— be it the pressure of air or gravity or something else altogether. It is taking the time to look down at the ground and up at the stars and understand that we are barely here in the universe, that our planet is an infinitesimal speck in what is likely an infinite expanse, to say nothing of our individual selves.

If our species manages to survive for more than another few thousand years, we will likely have to find a new home, a new planet as well as a means to travel and survive there. And this has to happen before the sun gets so hot that the oceans boil, some 500 million years from now.

But it is doubtful that even with our ingenuity we will be around for that long. After all, the average lifetime of a species is around 1 million years, although some manage to survive for as long as 10 million years. Our kind, homo, has been around for perhaps 2 to 3 million years. We ourselves have existed for perhaps 200,000 to 300,000 years. This is the middle-age of our species.

If by chance we breach the upper limit, it will probably be as a new, successor species that is unlike anything we are today. As much as science fiction may lead us to believe in so-called transhumanism, it is unlikely that consciousness exists as something separate from our bodies. It is highly improbable that we can upload and then download our consciousness at will, that we can transcend mortality.

We, like all things, will end. The light of human consciousness will be extinguished. Humanity will no longer be the default.

14

The Big Picture

We live under an illusion of homogeneity, of perfection created by distance. Like a pastoral scene viewed from the air or in a colour-corrected, soft focus digital photograph. It looks idealistic until you zoom in and see the blight on the landscape, the bare patches, the discoloration, the imperfections of the land.

But if you zoom out again, in time and space, we can start putting things in context.

We are 13.8 billion years into a universal expansion of spacetime that began when time itself began, at least in our universe and most definitely for us. But of course, we haven't been around for most of that time.

It is natural to ask what came before, and exactly what the universe was made from. But priors are always a problem because you could ask that question endlessly: and what came before that? and before that? and before that? Infinite regress. Nonetheless, we have something like a scientific explanation as far as our universe is concerned and that is this: nothing. Nothing came before. Or more precisely the question is meaningless and completely uninformative.

As far as we know, spacetime itself began then, with the big bang, with a random quantum fluctuation that generated an infinitesimal point of infinite density, what is commonly known as a singularity— like ones that exist in black holes. Of course, this does not mean that there weren't other universes, prior, parallel ones; or even ones birthed each time a choice is made, each time probabilities diverge— an endless series of multiverses forming continuously, ad infinitum.

We don't know what happened for the first 10^{-30} seconds after the big bang, a period known as Planck time. But shortly afterwards, there was, it seems, a rapid, sudden expansion, known as inflation, from 10^{-36} to 10^{-33} seconds, which disrupted the uniformity of the early universe and made matter possible— and ultimately everything we see around us— galaxies, stars, planets and us.

The first stars likely formed about a 100 million years after the big bang (that is, around 13.6 billion years ago), achieving nuclear ignition with hydrogen and beginning the process of manufacturing all the elements in the periodic table. The first galaxies would have emerged shortly thereafter, cosmically speaking. Our own Milky Way galaxy appears to have been among the earliest, with its first star born some 14 billion years ago (plus or minus 800 million years). While this pushes up against the age of the universe itself, the discrepancy is due to errors inherent in the various methods used to determine stellar age.

Our own sun, at about 4.6 billion years, is a sapling by comparison. The Earth formed about 4.5 billion years ago from the accretion of cosmic dust over perhaps a hundred million years that produced the planet we know today, although it has undergone many transformations since. One of those transformations likely included a collision with another planet that broke off a chunk of the Earth to form the moon.

As we look out from our little corner of the universe, we see an ever-accelerating expansion driven by an unknown force, referred to as dark energy. Everywhere we look, galaxies are rushing away from us at faster and faster speeds. Thus, calling this our corner is only useful as a metaphorical device— there is no true corner of the universe, there is no up, down or any direction at all.

We can only see and measure the observable universe, which is 92 billion years across (that is, we can see 46 billion light years in any given direction). Given the rapid expansion, we may never see what lies beyond the so-called "particle horizon," as it will take light from those parts longer and longer to reach us—if it ever does. And the reason that the observable universe is larger than it is old is because it is spacetime itself that is expanding, stretching distances between any two points in the universe.

And amidst this vastness, this unimaginable expanse, we have yet to find other intelligent life. Given the age of the universe and the number of stars we know there are in it, it is puzzling that we haven't met any aliens yet— at least not any that are verifiably and reliably alien. This contradiction is called the Fermi paradox, after the physicist Enrico Fermi, who first speculated about it in the 1950s. Since then, there have been numerous attempts to quantify the likely number of alien civilizations in our universe using the so-called Drake equation, which factors in probabilities for everything from the number of planets that could support life, to the likelihood of life emerging on those planets and the probability that life would have evolved to resemble eukaryotic multicellular organisms, which exhibit signs of both consciousness and intelligence.

On the one hand, it seems remarkable that there is any intelligent life, at all, in the universe. Even our own existence seems to be the result of uncanny luck. We live in a universe where the distribution of matter was just uneven enough for stars and galaxies to form, which in turn kick-started the cosmic, geological, chemical, and ultimately biological processes that led to us. Our planet also appears to be in the right neighbourhood, that is the right distance from the sun— not too hot, not too cold— to support life. These all seem like "Goldilocks" conditions, after the fairy tale of Goldilocks and the three bears, wherein the little bear's food, chair and bed were just right for the protagonist.

We could argue that in an infinite universe all things are possible and thus it was inevitable that such conditions would exist somewhere within it. But another way of looking at it is to say that we are simply suited to the conditions in which we find ourselves and it is equally likely that life elsewhere in the universe emerged to fit its particular environment— not unlike how evolution has worked for all these billions of years.

In 1974, the Australian Physicist, Brandon Carter, articulated the Anthropic Principle as a methodological argument for why the universe is as it is. In effect, he said that the fundamental constants of nature have the values they do (and thus the universe is the way it is) because if the values were different, we wouldn't be around to measure them. In other words, our presence as observers in the universe is only compatible with the conditions that we find ourselves in.

Even if so, perhaps we may be thinking too narrowly in assuming that life could only take the form that we find it in here on Earth. Perhaps we need to seriously consider whether other combinations of chemical elements— perhaps ones not even based on carbon—could lead to life, even to types of intelligence that we cannot fathom. Of course, science fiction authors have been imagining these possibilities for generations and who is to say that they aren't onto something?

Whatever form life takes, we can confidently say that it wouldn't exist without stars. This if of course because we and everything around us is made of star stuff. It is in these giant cauldrons that the elements of the periodic table are literally forged through the process of nuclear fusion. The gases that power stars, hot as they are, inevitably escape the stars' cores, as mass turns to energy, and leaks into space. And more dramatically, stars sometimes collapse under the weight of so much gas and go supernova. The elements they manufactured end up strewn across space and, from time to time, get caught in the orbit of some other star. Then, under its gravitational tutelage, those elements accrete into something resembling a planet.

That planet, becomes its own chemical factory, influenced in part by its position relative to its star, its orbit and by the particles that continue to stream from its sun and outer space.

On that planet, then, random collisions create a soup where more chance meetings result in lasting bonds among elements ultimately give rise to the precursors of life— amino acids, DNA, proteins, microorganisms, animals, people. Us.

15

When are We

Time is something we experience. We feel its passing. We have a sense of the seconds ticking. It is, if nothing else, an internal measure of change. But there are moments when our sense of it gets distorted, as when we are unconscious, asleep, intensely focused or distracted. When asleep or under the influence of anesthetic, we sometimes experience timelessness, yet when awoken have no idea of how much time has passed or even if it has passed at all. Time is what seems to order our lives. It helps us catalogue and arrange our memories, allows us to look back, learn and plan for the future.

Yet it can also seem very abstract, hardly graspable. Why for instance, does the arrow of time only go one way—forwards, rather than backwards? Could it have parallel paths? For all its ubiquity, for its seeming inevitability, time does not arise from the fundamental laws of physics, nor does it seem necessary for physics to be true or for the laws to work. Einstein's former professor, Herman Minkowski, unified space and time as spacetime, characterizing time as just another dimension. Spacetime is the fabric on which the universe is knit. It is what expands, it is what arises from the universe's gravitational field.

Nevertheless, time is illuminating on its own. Physicists think of time more in terms of entropy than in seconds. They characterize the flow of time as the transition of systems from states of order to disorder. Consider how we think about and observe the universe. The universe's past seems more disordered than the present, as if chaos has given way to a kind of stability. However, even order and disorder are terms subject to debate, as what they represent—that is, what it means for a system to be "ordered"—depends on the definition of order, which may well vary. Nevertheless, systems observed in the universe appear to move from states of disorder towards thermodynamic equilibrium, an even distribution, a sameness based on the dissipation of heat.

But such changes could be happening in multiple dimensions, ones we are neither aware of nor can detect or measure. Indeed, could time simply be a trick of the mind, the brain's solution for keeping itself organized sufficiently to act in the world? Could it be that time is not a part of any objective reality, assuming of course that objective reality exists? Could it be that we are suffering from "presentism," the notion that the past, present and future are different, and we can only directly experience the present moment, the now.

What if instead, we subscribed to "eternalism," the idea that the past, present and future are simultaneous and real at the same time, that they can be experienced and measured in a direct way within a "block universe." This is a block of spacetime which contains everything we see around us. What if it were something along the lines of what the science fiction writer Kurt Vonnegut has described in several of his novels. What if we could be an observer standing outside of time, like someone in space with a powerful enough telescope that could simultaneously see not only multiple events unfolding on different parts of the earth but also events across all of time?

And perhaps time is personal and local but not universal. Perhaps, it is analogous to the Earth being locally flat— we don't feel or perceive its curvature day-to-day or feel its rotation— but globally spherical.

We also don't see white light as a combination of red, green and blue, only as something uniform and unitary. Thus, is our experience of time determined by the limitation of our perception and cognitive capacity? Whether it exists solely in the brain or also outside of it, we only know what we know, what we see, experience, can measure. I measure, therefore I exist, as it were. Our measurements tell us that time is influenced by both gravity and speed of travel, as originally posited by Einstein's theories of relativity.

Gravity slows down time, which is why clocks are slower closer to the earth than they are high above it. It is why GPS satellites orbiting above the planet must be adjusted every so often to keep them in sync with clocks on the ground. The clocks aboard these satellites lose, on average, about 7 microseconds per day due to their velocity and gain 45 microseconds due to gravity, effectively making them about 38 microseconds faster than clocks on earth.

Our perception of time is affected by the so-called light cone, which is defined by the distance that light can travel in all directions within a given time. We can only perceive events within the light cone. For all intents and purposes, everything outside of it is in the future, at least our future, although in an absolute sense it is in the present for an observer at the light's point of origin.

And even for that observer, it would take something on the order of pico- or nanoseconds for light to enter the retina and trigger an electrical signal that travels along the optic nerve and for that signal to go through several stages of processing before we are consciously aware of it. That is to say nothing of the possibility that we do not perceive the signal at all, only our expectation of it, as determined by the brain's model of its surroundings.

Thus, to speak of any kind of perception as being in real time is meaningless. Everything we perceive and are aware of is from the past, we can never know the true present, the absolute now. For events that are even more removed from now, we rely on memories, which reconstruct the past and provide a portal to travel backwards in time. Similarly, we travel forwards in time when we make plans, when we imagine what the future will be like.

But ultimately, time, like many aspects of what we think of as reality, is personal. It is filtered through our emotions, our thoughts, our experiences of the world around us. That is why time can seem to go slowly when we are doing something we don't enjoy and yet flies by when we are having fun.

Even beyond that, time appears to be important for brain function, for us to be able to compute, make decisions. It may well be a precondition for our cognitive abilities and perhaps even for consciousness itself. After all, what would consciousness be like if it was an endless present where nothing changed and all we experienced was a persistent awareness of now?

William James famously separated the I from the me— one independent observer, one self, for now and another for the experience of the self through memory. The "I" was the self as subject, the "me" the self as object. No matter the distinction, both rely on time for their very existence, for the experiences that they represent. In effect, then, there is no self without time, neither past nor future, no change. That seems a lot like death, the obliteration of the self or at least our personal experience of it.

We often talk about brighter futures, as if progress is inevitable. As if it has a relentless momentum. As if time marches inevitably towards it. But the future is not ours. It doesn't care about us. The future is simply the unfolding of chaos. It is entropy itself that is inevitable . . . the gradual but unstoppable cooling that leads to heat death (cosmically speaking). The end is a uniformity, a sameness that's spread everywhere until anywhere is as hot as anywhere else.

This seems to be our ultimate future— or the universe's since it is unlikely that we will be alive then— in about ten quattuortrigintillion (10^{105}) years. At least until another random quantum fluctuation starts a new universe.

In the meantime, here on Earth— here, in our heads— we have illusory cycles that delude us into assuming that history has a shape and that we can divine it, if only we tried hard enough. We even imagine influencing those cycles— changing history, changing the future by a mere act of will. But the future's not ours to see (que sera, sera!). It's not even ours to create. It is its own creature, or more precisely the delta between one state and other. Time is just a term for that difference.

Still, we can't let it go. We want a clear view of where we are going. We invest in certainty by blindly embracing words like goals, objectives, vision— all of which are meaningless if not defined with respect to time. We want to dream big, aim high; never settle. The future, then, has become a country where all our goals have been achieved, where our dreams have been fulfilled. It is a time when everything is not the same but better. Everyone tells us that we will get there—to this promised land. And there is no shortage of inspiration and affirmation to help us along the way.

Except "there" is nowhere, really. It's simply a waypoint, not the destination. The future is simply an idea— one that changes everyday. Sometimes, we get so attached to one version of it that we lose sight of where we are now or to where else we could go. In focusing too much on the destination, we forget to experience the here and now— and the journey from here and now to there and then.

For isn't that what life is— a journey that begins at birth and ends at death. It can and will take any number of routes to get from alpha to omega but that doesn't necessarily mean that one way is better than another. We could just as easily take the scenic route or not; the road less travelled or not; we could even stop and smell the roses or the cow dung.

And along the way, it may help to remember to get lost, every now and then.

16

Entangled Forever

Matter is what we are, star stuff and all. Energy is what we use and become. The two are interchangeable, as elegantly captured in Einstein's immortal equation, $E=mc^2$. And the total amount of matter and energy in the universe is a constant.

But as if matter, energy and time weren't strange enough, we have quantum mechanics, which describes the behaviour of nature's fundamental units— the so-called elementary particles— at the smallest scales imaginable.

Despite its oddities, quantum mechanics is a very successful theory in that experimental observations and evidence correspond to theoretical predictions. However, it is still subject to some degree of interpretation until more evidence can be gathered, in part through experiments like the ones being done with particle accelerators, which generate and then "smash" particles together to try and break them apart and discover their constituents. This is how the Higgs boson and various other elementary particles were discovered.

The most fascinating properties of these particles include, for instance, quantum superposition, which holds that the natural state of any particle is in a combination of all possible states, with distinct but unknown probabilities for each state. However, the particle's state cannot be resolved— it doesn't achieve what is called "decoherence"— until it has been measured.

There has been much speculation and fantastical theorizing about how and why decoherence occurs. Briefly, decoherence is the realization of a measurement, the identification of a particle's spin, its momentum and location. It is the point at which we gain knowledge about the particle's nature. One idea has been that consciousness has something to do with the "collapse" of the quantum wave function, which represents the probability distribution of the particle's possible states.

Specifically, the theory goes, probabilities resolve upon decoherence, and one state becomes real only because it was consciously perceived. In other words, if a tree falls in the forest and no one is around to hear it, it doesn't make a sound— or perhaps it never falls in the first place, it exists in a superposition of all the angles between being upright and lying on the ground.

If this is true, then what does it mean for us— our bodies, our selves, even our life as lived in time. Are we constantly in a state of flux, embodying a multitude of possibilities until we are measured? And who does the measurement— another observer? Or us, through our own individual, subjective, and conscious experience. If the former, then why is it that we experience a concrete reality, a tangible flow of time— a single, definitive experience from one moment to the next— rather than a superposition of all possible experiences in any given moment.

In some ways, this is analogous to Heisenberg's uncertainty principle, which states that the position and momentum of a particle cannot be measured at the same time, since taking a measurement of one property changes the other. This is often translated into lay terms as, "to observe is to disturb." Thus, it is not possible to measure a thing without affecting it in some way. The very act of observation changes that which is being observed to such an extent that it becomes practically impossible to know the nature of objective reality (the world as it exists without an observer). This is commonly known as the "measurement problem" in physics.

Luckily, there appears to be a quantum-classical transition that renders us fully resolved, decoheres our respective wave functions, which makes us real, in a sense. This quantum-classical transition may perhaps be a case of imprecise measurement. Without sophisticated instruments, we simply cannot measure the decoherence inherent in us— we can only resolve one of many possible worlds— namely, the one we live in. What we experience, then, are the results of our own conscious measurements, which are much fuzzier, imprecise, and incomplete. It is like seeing a giant, seemingly uniform blob floating in the distance that resolves into a swarm of millions of bees when we get closer.

This may partly be because we ourselves—the particles in our bodies, the signals coursing through our neurons—are all part of the wave function, which can therefore not maintain its coherence in us. And the many worlds interpretation of quantum mechanics— posited by the physicist Hugh Everett in 1957 and still hotly debated— suggests that all possible histories continue to exist— the wave function maintains coherence in parallel universes. This means that there is no decoherence, no wave function collapse. It is simply that we experience only one possible manifestation of the universal wave function.

Putting this in terms of the infamous Schrödinger's cat thought experiment, before the box is opened, the cat is both alive and dead. When it is opened, it is either alive or dead— only one possibility is realized: only one reality manifests. Under the many worlds interpretation of quantum mechanics, the cat is perceived as alive in one universe (say ours) but is dead in another one. Thus, the wave function maintains coherence and all possibilities remain viable.

This is a staggering idea, with mind-boggling implications, one being that there are infinite histories and therefore infinite universes containing every imaginable realization of every event that has multiple potential outcomes. Each moment generates new possibilities and therefore new histories. How fine that slice is, remains an open question because the answer depends on how divisible time itself is.

But the wonders of the universe are many and at its smallest scales, there is another one that presents its own set of mysteries. This is the phenomenon of quantum entanglement, whereby two "entangled" particles appear to continue communicating in some manner— and do so at speeds faster than light— even when separated in space.

Entanglement arises when a pair or group of particles are close enough to interact with each other. Their various properties, including, for instance, spin, momentum, polarization, and position can be measured. Then the particles are separated and dispersed to different points in space. At this point, if the properties of one of the pair is measured, it is possible to immediately know the properties of the other— they seem to be perfectly correlated across distances and speeds that would have been impossible even with information travelling at the speed of light from the measured particle to the yet to be measured particle.

There are competing ideas about exactly what is going on, including questions about whether any information is actually being exchanged or if there is something about entanglement that primes the wave function in a manner that enables correlation. But beyond this, there is the question of whether the observer is actually having an effect on the outcome of the measurement.

The laws of quantum mechanics suggest that it does not even make sense to talk about the position of a particle, that it is not a property that particles have. This is because detectors and observers are all entangled— they form part of the same quantum mechanical wave function. However, we are unable to perceive this, just as we are unable to perceive the motion of the Earth even though it is spinning on its own axis, rotating around the sun, and moving through the galaxy— all at the same time.

While quantum entanglement continues to accrue more and more experimental evidence and even enable practical applications such as cryptography and quantum computing, it is perhaps more intriguing in terms of its implications for us, personally, and for the environment that gives us existence.

17

All That We Are

Life on Earth is thought to have begun some 3.5 to 4 billion years ago. Exactly how it started though remains an unsolved mystery. Did it, as is posited by the so-called panspermia theory, already exist in some form elsewhere and travel to Earth aboard meteorites that rained down on the planet during what is known as the late heavy bombardment between 3.8 and 4 billion years ago? Could it be that at the very least, those meteorites contained amino acids, the precursors to life, which kickstarted the process on earth?

Certainly, there is now evidence of amino acids in comets, as well as in meteorites that have fallen to Earth more recently. But it is also possible that the primordial oceans had the right conditions, the soup of chemical elements necessary for nucleotides to form organically and eventually assemble themselves into DNA. Some experiments, starting in the early 1950s (Stanley Miller and Harold Urey) have demonstrated that organic molecules can self-assemble from inorganic ones, under the right conditions, and subsequently self-replicate. However, this remains an area of active research.

Regardless of how it emerged, life seems to have begun with single cell organisms that over time adapted to their environment through a process of mutation, combined with natural selection that led to more and more complex forms.

For the first billion years or so, there was nothing more sophisticated than bacteria. Then, based on the fossil record, some 530 million years ago there was a sudden (in geological time, at least) diversification life, referred to as the Cambrian explosion. Since that time, the planet has been through various climate shifts, including changes to the configuration of the land mass and several ice ages.

The last ice age was some 12,000 years ago and since then we have been living in an epoch termed the Holocene, although some are arguing for a declaration— possibly going back several hundred years— that we have entered a new age, the Anthropocene, or the age of humans. In 2024, a panel of experts declared that a new epoch had not yet begun.

Dissenters argue for declaring the Anthropocene based on the argument that humans have significantly altered the trajectory of the Earth and its climate, extracting its resources, destroying natural habitats, and releasing carbon into the atmosphere much faster and on a much larger scale than would have happened naturally.

Despite this significant impact, we still have something on the order of 8 million different species that have been identified and many more likely yet to be discovered. And all these species have been changing as the environment changes.

Each is sophisticated in its own way— even a single cell organism. Thus, none can be said to be primitive, especially not after surviving for hundreds of million, or even billions of years in the case of archaea and bacteria— remarkable evidence of resilience.

Nevertheless, most species that are alive today have only been around for a few million years. We ourselves are a tiny twig on an intricate and many branched evolutionary tree, and a recent addition at that. Yet here we are putting forward our intelligence as unique and unprecedented in the history of the planet. And even if it were, by no means do we represent an inevitable culmination of a process of perfection via evolution.

If anything, every species that is alive today has succeeded and is fit, for the moment, to exist and perhaps even thrive. And that is a form of intelligence—the ability to sense and respond to the environment in a way that ensures survival, especially survival long enough to ensure reproductive success.

If we ourselves have no master controller, no homunculus exerting the will that drives our beings, much of what we are today is the result of a simple adaptive, deep learning biological algorithm playing out to no particular end.

If true, then all living things lie on a consciousness and intelligence spectrum, with mostly just enough of it to adapt to a given environment. Despite ongoing philosophical and scientific debates, a consensus of sorts has emerged, with the Cambridge declaration that states, in part:

"Non-human animals have the neuroanatomical, neurochemical, and neurophysiological substrates of conscious states along with the capacity to exhibit intentional behaviors. Consequently, the weight of evidence indicates that humans are not unique in possessing the neurological substrates that generate consciousness. Non-human animals, including all mammals and birds, and many other creatures, including octopuses, also possess these neurological substrates."

And this has been borne out through many experiments. Take for instance, apes, who, while not exactly a match for humans, are still thought to have theory of mind. That is, any given ape demonstrates the ability to think about what another ape is thinking, perhaps even what another animal (including a human) may be thinking or planning; and further, they seem to be capable of making decisions based on such inferences.

Further, bonobo societies exhibit complex language skills based on facial expressions and have been seen taking care of the old and disabled; have been known to adopt orphans; and, like humans, violently pursue power. The primatologist, Frans de Waal, has recorded multiple instances of apes demonstrating empathy, even for other species. In one classic example, he noted how a bonobo named Kuni, which was being held at a British zoo, handled a bird that had collided with the ape's glass enclosure. Kuni, picked up the bird, climbed up to the top of the tallest branch within the enclosure, unfolded its wings and let it fly away.

But we need not stop at animals, some neurobiologists believe that plants, too, are conscious. However, they may not be self-aware, with the ability to know that that they are conscious or contemplate their consciousness in any meaningful way. Nonetheless, plants have a sense of their existence in space and perhaps also in time and have evolved ingenious solutions to being fixed in place and survive their so-called "sessile" existence.

Part of this involves sensing the environment to collect data, in as much as human perception allows us to understand and navigate our world. But plants seem to have not only achieved parity with our five senses (that is, they have equivalents of touch, smell, hearing, sight, and taste) but to have gone beyond us. They have almost 20 different senses, depending on the species.

Plants' senses are attuned to their needs, including knowing how to orient their leaves; when to open a flower or unfurl other parts; when to produce chemicals based on the presence of predators, which are detected through chemical or sound signals; and how to find water and other nutrients in soil and navigate around or through obstacles, as roots often have to.

Roots are especially well adapted, with many having developed the ability to respond not only to the presence of soil nutrients and moisture but also to pressure, gravity, volume, microbes, density and much more. There is also evidence that suggests plants communicate with each other, fungi, and animals through chemical signals; and that they provide nutrients to weaker plants.

For instance, maize release a spray of chemicals that attract wasps to lay eggs in caterpillars, which munch on the maize— clearly a sophisticated response to a threat and evidence of evolutionary adaptation.

Another example is of how some plants change the texture or flavour of their leaves based on signals received from other plants that have detected the presence of an herbivore, chemically analyzed its saliva and determined the appropriate response to counter this threat. They can change the toxicity of the leaves enough to kill the predator.

Plants also appear to communicate through their root systems and exchange chemical signals with other plants and possibly other species, warning them of dangers, reporting on environmental conditions and so on. If supported by further research, this would be quite astounding and should force us to reconsider how we view plants, or perhaps even our use of terms such as "vegetable" or "vegetative state" to describe those that have been incapacitated or rendered unconscious. This is especially so because plants represent something like 99% of the biomass of the planet.

Despite this variety of seemingly intelligent behaviour, plants clearly lack any type of brain or nervous system that we can discern. Nonetheless, there is evidence of plants' ability to learn and form memories and change their response based on what they have learned.

This has been demonstrated in experiments with a plant called *mimosa pudicas* (more commonly known as touch-me-not or the sensitive/shy plant), which curls its leaves on touch. When experimenters dropped these plants, they sensed the fall and curled their leaves in defense. However, after several attempts, the mimosa seems to have concluded that no harm was imminent and stopped curling their leaves when dropped.

In addition to their remarkable and long unknown intelligence, plants are highly modular— they can regenerate from less than 10% of their form. Nonetheless, the idea that plants have the equivalent of a brain remains controversial though there are biologists who believe that this cannot be ruled out— even in the absence of anything that looks like neurons.

On the other end of the spectrum are microorganisms and even larger organisms such as jelly fish, that don't have a brain or nervous system. Jelly fish are an interesting case because many species can't even control their own movements—they just drift with the currents.

Evolutionarily, jellyfish are among the oldest living things and yet their genome has much in common with our own. One species, *turritopsis dohrnii*, could even lay claim to being immortal because it can return to its embryonic polyp stage and regenerate— again and again— in response to damaged tissue or toxic environmental conditions.

It does so through a process called transdifferentiation, which follows a path similar to so-called pluripotent cells (like stem cells) that have the ability to turn into many different types of cells. In the case of jelly fish, however, it is the cells that are really immortal, not the organism per se.

Closer to the root of the evolutionary tree, even the simplest fungi appear to have evolved remarkable abilities that in some cases exceed our own— at least when not considering tools such as computers. For instance, slime molds can find the most efficient route between two points regardless of the number of obstacles in their way. In effect they can design complex transit routes that we would not be able to chart without the help of computers.

All this raises important philosophical questions. For instance, why do we believe that humans are special, that our lives are worth saving more than that of other living things? And vegetarians and vegans also have to consider why they believe it is better to eat plants rather than something like jellyfish, or any living thing at all.

The answers are not straightforward— we cannot rely on existing moral standards, whether derived from religion or elsewhere. And first, we must acknowledge that we are part of life, not removed from it and permitted to exert our will on other forms of it without somehow affecting ourselves. Life on Earth exists within an ecosystem and humans are but a part of it—although we may indeed be unique in terms of both extent of self-awareness and cognitive capacity.

18

Being in the World

The world is obviously here. We see, hear, smell, taste and feel it. We do things in it; we have experiences in it. It does not seem to be an illusion of our predictive brains. Yet, it is a strange place, wondrous even. We do not entirely understand its working, especially at the very small and very large scales. We do not fully understand what it is made of and by extension, what we are made of. We wonder if our thoughts are different from the physical stuff of the universe or if they are simply a property of matter. We wonder if we can trust that anyone else, other than us, is conscious or real, since we can only access our own experience.

But could the entirety of our world be a shared hallucination, with belief and consensus as the knit that holds it together?

Out there in the world, we are performing and all of us are actors on the world stage, however much we pretend that we are not. In fact, we go to great lengths not to bring down the third wall between us and our audience, even when that audience is us.

Think of the effort we expend to hide the messy elements of a tv, radio and sometimes even a stage production through clever editing, computer graphics, green screens and makeup. We are complicit in a charade— everyone knows what's beyond the edge of the screen, what's been cut out and papered over, where the wrinkled have been ironed. But if we pretend long and hard enough, we forget that we are pretending. We start to believe that the performance is reality.

Indeed, our entire social existence is a construction. In their 1966 book, "The Social Construction of Reality," Peter Berger and Thomas Luckmann alluded to the likelihood of our social reality existing by consensus. Things are only as they are if we all agree on what "common sense" entails and then follow through on the rules derived from it.

We agree to stop at a red light and go on a green one. It is that tacit assumption and acceptance of what everyone else knows, which allows us to live a somewhat predictable and stable existence. In its absence, we would quickly descend into anarchy and chaos. Everyday, we are unconsciously yet actively creating order out of disorder, creating reality out of meaninglessness, because the opposite is unfathomable— because the only other option is social breakdown and existential terror.

But being in the world goes beyond our existence and performance in human society, beyond our attempts to imagine and create social reality. We live, after all, in the natural world and are subject to its forces. We interact with other living and non-living things, and we depend on these for our survival. The resources we extract from the world support our modern lifestyle. So, it makes no sense to believe ourselves as being somehow independent of it.

There is a tendency to think of humanity as exceptional, given the supposed mastery of our environment, our ability to adapt to it, as well as our track record of adapting the environment to suit us. This human exceptionalism often leads us to believe that we have escaped biology, that cultural factors now play a bigger role in our evolution and will ultimately help us escape organic constraints.

The case for such thinking looks robust. For instance, the most recent evidence of evolutionary change in humans includes our ability to continue digesting lactose after weaning— even though there are still large numbers of people with lactose intolerance. The explanation for this new ability is that animal domestication, which began ten to fifteen thousand years ago, led to humans co-evolving with cows, goats and the like, thereby gaining the ability to digest their milk.

More recently, with genetic engineering, prosthetics, laser eye surgery, cochlear implants, retinal transplants, and numerous other innovations, humans seem to be gaining an edge over biology. Our ability to repair, change and enhance our bodies is unmatched by any other living thing on the planet. Does this mean that we are free of natural constraints? That we are no longer subject to the so-called natural laws or indeed the laws of physics?

That depends. At this point the debate becomes more philosophical than scientific, but we could make the following argument: The universe is subject to the laws of physics and all matter, including humans, are subject to those laws. Biology emerges from physics and chemistry and is therefore subject to the underlying laws discovered in those disciplines. Consequently, humans can be no freer of biological laws than they can be free of physical or chemical ones.

Clearly, however, we can use those laws to our benefit. We can deduce the chemical structure of DNA and manipulate it to correct malfunctioning cellular machinery, for instance. But it gets more complicated when we start talking about whether we can similarly manipulate human behaviour— not by using the underlying laws— but through culture alone.

There is no conclusive or satisfactory answer right now. But without getting stuck on that point, we can perhaps agree that it is important to recognize that we are of the natural world and not separate from it. And further, agree that we can only survive in an ecosystem— we are unlikely to be able to exist entirely in a bubble of our own making.

However, if some of the most ambitious space colonization plans come to fruition, we may get to test this idea of self-sufficiency by establishing human settlements on distant worlds like the moon or Mars and surviving there indefinitely and independently of supports shipped from Earth, including food, water, and energy. It would entail creating a whole new ecosystem on another world, which would also need to support the trillions of organisms that live on and in us and are essential to our survival.

In the meantime, most of us alive today are stuck on Earth and in our current ecosystem. We must co-exist with all the other species on the planet. And while we are here, we are learning more and more about how many aspects of ourselves that we believed to be unique are not so. Other animals also exhibit remarkable levels of sociality, including cooperation, caring for kin, planning, and theory of mind, possibly even varying degrees of self-awareness.

It helps to understand where we have come from, biologically, to determine if we are really so different from other species—if we are indeed as exceptional as we think we are. All indications point to humans as a relatively recent branch of the so-called tree of life, which is perhaps more of an intricate web not only in terms of evolutionary relationships but also in how we survive today, within a larger ecosystem and the food chain.

We are, after all, the product of some 4 billion years of evolution, starting with simple, single-celled organisms (close relatives of which still exist) along with millions of other species. We have developed classification systems that organize all known life into so-called kingdoms or domains, which include bacteria, archaea and eukarya, although some scientists argue for the recognition of a fourth domain that would include viruses, which are still considered to be suspended between the animate (living) and inanimate world.

We ourselves fall into the eukaryotic domain (organisms whose cells include a nucleus, which encapsulates DNA that likely emerged from the improbable yet fortuitous encounter between archaea and bacteria, about 1.5 billion years ago). We also evolved from fish, as did all mammals and amphibians. But our commonality runs much deeper and broader.

For instance, the molecular precursors of learning and memory appear to stretch back over 3.8 billion years to bacteria. And the present-day molecules that enable brain function are shared among both vertebrates (like us mammals) and invertebrates (e.g., insects) and can be found in our last common bilateral ancestor (that is, animals with symmetric bodies) from 600 to 700 million years ago. And although these are organisms that did not have a nervous system, the molecules they evolved clearly presaged the development of more complex brains.

In fact, many of the mechanisms of life—from the ability to metabolize energy by finding and processing nutrients in the environment to sensing and responding to threats and maintaining internal chemistry—are shared. The extent of commonality in the foundational elements of life is staggering. At the most basic level, our current definition of a living thing entails that thing being made of a cell— one or many— and of there being DNA in that cell— either contained in a nucleus or free floating in the cytoplasm.

While there continues to be a debate over whether viruses are alive, they do possess genetic material (it is typically RNA, not DNA; and there is a class of viruses that do have DNA). RNA viruses cannot replicate on their own, rather they need to enter a host cell and then use its genetic replication machinery to reproduce.

Regardless of the specific status of viruses, their existence, together with that of all other organisms on the tree of life, is overwhelming evidence of nucleic acid as the shared basis of life. The same four chemicals (adenine, guanine, cytosine, and thymine or A-G-C-T) arranged in different sequences are responsible for every single living thing on the planet and for the tremendous diversity of it.

Further, there is just a 0.5% genetic difference between each human. And not as much difference between us and other animals. Consider that we share 96% of our DNA with chimps and 60% of our DNA with mice. We also have a surprising number of characteristics in common with fish, including lungs and the groove that runs between our nose and top lip. Our hands and feet, fingers and toes can be traced to the same genes that code for fins. Indeed, *tiktaalik*, an intermediary between fish and amphibians that lived in mud pools almost 400 million years ago is one of our ancestors.

Our last common ancestor with apes is thought to have existed about 10 million years ago and our lineage of relatively modern humans may go as far back as 2 million years, although our ancestors likely started walking upright as much as 3.7 million years ago. Despite all this evidence, a clear picture of our immediate forebears has yet to emerge, including the degree to which our species, homo sapiens, may have bred with and replaced other homo species, such as the Neanderthals, Homo Nadeli, Homo Erectus, etc.

But we didn't really start down the path to become who we are until the development of rudimentary social intelligence, likely starting with Australopithecus, a related species of hominins that lived 2 to 4 million years ago. Compare this with ants and termites, which evolved eusociality 150 to 200 million years ago.

Eusociality, which is unique to insects, entails building shelters to protect the young and rearing them to maturity before they are allowed to explore outside the nest. Meanwhile, adult members of the group, including unrelated ones, cooperate to raise the young, by searching for, finding, and transporting food back to the nest. This obviously requires a significant amount of coordination, cooperation, and communication.

While we did not become eusocial, our forbears did increasingly start to cooperate, likely driven by the need to meet the greater energy demands that arise from being endotherms, or warm-blooded animals. At the same time, the theory goes, Australopithecines switched from a vegetarian to a meat diet, which while providing more calories, was also more difficult to obtain. Thus, social behaviour may have presented a unique advantage, since those who cooperated had better access to the more energy-rich meat diet. The side effect of this denser diet was a significant increase in brain volume from 500 cm3 to 600 cm3 in Homo Habilis, 900 cm3 in Homo Erectus and 1,400 cm3 in Homo Sapiens.

All this appears to have happened in Africa before we started migrating off the continent to practically every part of the planet, which by the most current estimates may have started anywhere between 170,000 to 150,000 years ago. It likely began as a land migration (including over the Bering straits to the Americas some 20,000 or more years ago) and, over many thousands of years, broadened to include travel by sea, especially for the hop to Polynesia and Australia about 60,000 or 70,000 years ago.

There may have been two main initial routes, a northern one through the Levant (eastern Mediterranean region that includes modern Lebanon, Syria, Israel, Palestine and parts of Turkey) and a southern one via Ethiopia and the Arabian Peninsula. During the last ice age, which lasted from 80,000 years ago to 11,000 years ago, both the Bab el-mandeb strait near modern day Ethiopia and the Suez basin near modern day Egypt were shallow and dry, respectively. This would have allowed for the migration out of Africa, not just of our own species but also of related ones, such as Neanderthals and Denisovans, who started diverging from each other some 400,000 years ago.

But much of this is still being debated and there has been a steady flow of new discoveries that have led to revisions of the timeline and to the evolutionary path from Homo Nadeli, the earliest homo species, to Homo Sapiens, latter day humans. For instance, discoveries in 2018 suggested that there may have been a homo species in China as long ago as 2.1 million years, although this was not Homo Sapiens. Therefore, the story of our recent evolution has by no means been fully written.

Nevertheless, the consensus about modern humans is that we likely emerged in our current form about 300,000 to 200,000 years ago. For most of that time, we have been wandering the planet, much as other animals do, though perhaps more extensively, covering greater and greater swathes of the Earth as we built boats and then other rudimentary forms of transportation to take us to every continent on the planet.

The wandering lasted many hundreds of generations. We only started to settle down 15,000 to 10,000 years ago with the invention of agriculture and animal domestication. This subsequently led to the development of cities and to the beginnings of civilization as we know it today, around 6,000 years ago.

Thus, on both cosmic and geological scales, our species is young and our modern age younger still, at less than a couple of hundred years old. In fact, if we were to compress Earth's history into a year, all human history would fit into the final day of the year. At the cosmic level, our entire existence would account for about a minute in the universe's 13.8-billion-year history and an individual life would barely account for two-tenths of a second.

Despite this brief period, we appear to have grown to dominate our environment like no other species to ever inhabit the Earth. And now, we decry as naturalism any attempt to compare and equate humans to other life forms. We believe that we are unique, specifically because of consciousness and more so because of our self-awareness. Our subjectivity, our so-called shared intentionality— the ability to plan and follow though on a plan, together— is considered a distinctively human trait. So is our ability to express ideas together in complex ways. Even something as simple as pointing and episodic memory that we can traverse at will are all said to be characteristics that no other animal possesses.

What then is our place in the world, at a time when through our own ingenuity, we find ourselves at the top of the food chain. We likely consume more of Earth's resources than any other life form, but we are not as numerous. We cover or can reach practically every corner of the planet, but so do many microorganisms. Further, insects and microorganisms outnumber us. One person likely harbours more bacteria, viruses, and fungi than there are people on the planet.

And while we have taken many capabilities to the limit through knowledge of fundamental natural laws and feats of engineering, other organisms still do many things better than us— whether it is harvesting energy, or manufacturing enzymes and compounds. They also appear to form social bonds and cooperate as, for instance, the neurobiologist Peggy Mason has demonstrated with groups of rats in her lab. In the latter case, the rats demonstrated both emotional contagion (mimicking the emotional states of others) and a certain level of empathy, by attempting to liberate trapped fellow rats. Another species, the prairie voles also exhibit empathy for their bonded mates.

At the same time, we have co-evolved— with the microorganisms in our body (including many that have contributed to our genome) as well as with other animals. For instance, African birds called Honeyguides direct people to wild honeybee hives so that they can access bee wax and larvae once their human co-conspirators have smashed the hives.

The most astonishing part of this relationship is that the honeyguides and the honey hunters communicate through calls, with humans calling the birds when they are ready to begin, and birds responding to those calls. Further, birds in different regions respond to different calls, such as whistles in Tanzania and other sounds in Zambia and among the Yao people in Mozambique.

Our connection to our fellow living things is also evident in early development. In the 19th century, following Darwin's formalization of the theory of evolution, the German zoologist, Ernst Haeckel built on ideas that had been around since the late 18th century. He proposed that ontogeny recapitulates phylogeny— that is, our embryonic development mirrors and re-plays our evolutionary development. This theory has since been abandoned although biologists still note that embryos of different species resemble each other much more so than the adult equivalents, which may point to our shared evolutionary history— even if we don't exactly replay evolution in the womb.

19

Being Social

Why, as humans, do we congregate in groups? Why we do form attachments, both fleeting and lasting? Why do we live as families/tribes/nations? Why don't we, like many other animals, lead a largely solitary existence, meeting to mate then going our separate ways, hunting alone, eating alone, perhaps even dying alone?

Evolutionary biologists theorize that being social is fundamental to our survival as warm-blooded creatures that have high energy needs and are born relatively weak. We need groups to help protect our young and to help each other survive, especially because we don't have the physical prowess of other animals.

Of course, we are not the only creatures that cooperate. Sociality is found among apes, elephants, prairie voles and many other species. Similarly, insects exhibit remarkable degrees of eusociality, with unrelated members of the group caring for the young and cooperating to find and transport food; build and maintain habitats; and defend the group against predators.

It seems likely that there is an evolutionary advantage to cooperation and that the human form of it is not exceptional. And while cooperation is natural, the answer to the question of what we owe each other may well be "nothing," at least from a purely biological perspective. Nonetheless, we help each other; perhaps out of self-interest. Social groups allow us to establish trust with small numbers of people in different settings. With them we form a mutual confidence and supply agreement, as it were, to prop each other up, to help when help is needed and to mostly not harm each other or each other's interests.

This, then, is the context of our relationships, admittedly one stripped of sentimentality and our subjective experience of being in those relationships. The point though is that altruism, which has a personal cost and perhaps no immediate return or no return at all, is still beneficial. It is in our interest to work with each other, to find ways of living together so that our genes can persist through generations.

Day-to-day, it is much more complicated than that. We add layers of meaning based on inferences wrought by our ancient, emotional brains. We ascribe motives, we assume intent, we take offense, we get defensive. We hug. We fight. We love. We kill. And we do all this because often we cannot do otherwise, because sociocultural evolution hasn't trumped biological evolution to the point where we can engineer each others' behaviour.

We have irrational fears about others' reactions to us. While it may be justified by experience, in specific instances, it is irrational when there is no historical basis. We start with a baseline of anxieties developed from experience and project them onto new situations, and we make further assumptions and start constructing a narrative that takes on a life of its own.

As fraught as they are, we obviously need other people. Few of us can exist in isolation, even putting aside practical considerations. We start to feel lonely and, as we do, we may further isolate ourselves, which in turn could lead to depression and poor health and increased risk of mortality.

Thus, it is important to understand the context of our relationships and their nature because we are often led astray by what we think is going on with other people— the types, quality and quantity of relationships that they have. And when we compare, we may feel inadequate, as if we have failed to meet some social standard, been left behind, excluded.

Evolutionary biologist and anthropologist Robin Dunbar has posited an intriguing connection between a primate's neocortical volume (the volume of the brain's neocortex, which is responsible for a lot of so-called "higher order" capabilities or executive function) and the number of social connections it is able to maintain. His research suggests that humans can, on average, sustain about 150 associations, including family, friends, co-workers, and others.

There is anecdotal evidence to support the so-called Dunbar number, even though some have suggested that it may be higher, and others have argued against drawing any conclusions from the correlation between brain volume and social connections. For instance, the typical size of army companies is 150 and has been since Roman times. The company and similarly sized groups are then organized into smaller sets of 15 to 50 members, which corresponds to the size of many hunter-gatherer tribes.

Dunbar's number can be represented as a series of concentric circles that enumerate the cumulative number of relationships. The innermost circle includes 5 people— these are the family and friends that are closest to us, the next circle includes close confidants, which could be about 10 members. The third circle may include up to an additional 30 close friends. The circle after that includes up to 80 causal friends.

Overall, Dunbar contends that humans can keep track of around 500 acquaintances and can match, at most, about 1,500 names to faces. If true, this highlights the irrelevance of the number of connections someone may have on social media, which are ostensibly designed to keep us in touch with people that we supposedly know.

If someone has many hundreds or even thousands of connections, it is quite likely that they barely know most of those people or that the relationships have gone stale. Additionally, consider the effort it requires to maintain a relationship. Research from Jeffrey Hall and associates suggests that it takes at least 50 hours of interaction for two people to go from being acquaintances to becoming casual friends. A further 40 hours are required before those same two people become real friends. In total, the researchers estimate that at least 200 hours of socializing is necessary to form what most people would think of as close friendships.

And while it may not hurt for some of this socializing to happen remotely and / or asynchronously, via social media, for instance, face-to-face interactions are still important to maintain and strengthen relationships. Perhaps biologically our relationships are still mediated by the parts of our brains that process sight, smell, sound, and touch. Thus, we need to see people, hug them, laugh with them, and—consciously or not—smell them to draw closer; to maintain the ties that bind; make us willing to do things for each other, even when we don't feel like it; even when it may cost us financially, physically, or emotionally.

All the same, remote interactions mediated by social media are not altogether bad or useless. They could in fact help to keep alive some weaker relationships that may otherwise wither. Technology makes it easier to maintain such looser, possibly less valued, associations that we do not want to discard entirely.

Of course, in a culture that is focused on performing relationships, we have to wonder how many relationships are real and how many fall into the association category. We may be better off severing those looser and more distant ties, rather than holding on to them out of some misplaced hope. It may also be more compassionate to let those relationships go, rather than leading someone on, pretending to be their friend, sapping our energy and theirs. Instead, we could invest in the people we do want to hold onto.

This requires honesty, first and foremost. It means honestly admitting whom we care about and whom we don't. And perhaps it also requires us to go deeper, to figure out why we care about some people and don't care about others. Then we can decide whether it is appropriate to reinvigorate a flagging relationship— friendship or otherwise— or to let it go; to invest or to divest.

We also need to think more broadly about friendships, possibly by using Dunbar's numbers and classification as a guide. We can redefine friendship as a collective term that describes many different associations with various types of people— those with whom we would share emotions and intimate details; those we associate with because our children play together, go to the same school, engage in the same activities; those with whom we work; those we are friends with because our parents were friends with their parents; those with whom we go to book club meetings with or exercise with from time to time. Close friends, phone friends, casual, activity-based friends, parent friends and so on. Such a re-definition will help us see our relationships more clearly and allow us to recalibrate our expectations of others more intelligently, while allowing us to decide how much of ourselves we invest with whom.

Ultimately, there is no moral authority that can help us decide who is worthy of our time and attention and who is not. Philosophers already argue about the merits of utilitarianism or other ways of allocating limited resources. In a similar vein, the question is whether we should focus on a select few or on attending to as many people as possible, even if they are not all equally worthy of our attention. Less worthy, that is, not in any objective sense but by virtue of their characteristics and relationship to us (thinner, fatter, younger, older; friend, foe, family, stranger).

Biologically, we are primed to favour kin. Thus, it seems natural to begin our charity at home before we go off and try to save the world. Despite all the philosophizing, it may not be about benefiting the greatest number of people, or even the neediest members of our or other species. In fact, it may be more about taking care of us and ours. Admittedly, this sounds like nativism and perhaps it has an element of that. And it is possible if we weren't ultimately animals subject to biological determinism, we may do otherwise. But we can't entirely escape a nature designed by an effective but imperfect mechanism— evolution.

So, are we deceiving ourselves when we claim to care about distant peoples and their troubles; their wars, diseases and other suffering. Some attempt to alleviate such misfortune and may even spend whole lifetimes doing so. But are these impulses possibly less altruistic than they seem. Are they ultimately about feeding our ego than helping someone else. Reported abuses of the humanitarian system, and of those that it seeks to help seems to support that kind of conclusion. But at the same time, it is hard to deny that some benefit is actually realized.

Ultimately, it is less about whether saving the world is a good thing and more about where that falls in our personal list of priorities. Few of us, after all, would sacrifice our own children or loved ones for the sake of the world. So, perhaps it makes sense to start locally, to solve our problems at home—however defined—before trying to solve bigger, more remote problems; possibly even before trying to help those who seem to need the most help— to fulfill the dreams of the world's utilitarians.

As much as we have come to believe that there is some universal morality, some absolute measure of what's right and wrong, it is impossible to prove, unless we subscribe to one or more religious doctrines or secular ideology. Of course, this could be seen as a slippery slope— if we are to base our lives on what we understand of our biology, if we give into biological determinism then there is no point trying to improve most social ills— discrimination, mal-treatment, genocide. But should we be so quick to throw up our hands and walk away from the idea of equality and justice for all?

It is a conundrum, one fraught with contradiction and risk. That doesn't mean we should look for some objective truth to guide us towards the right thing. The right thing is what we agree it is. And even that decision is a complex one. How, after all, do we agree? Is consensus necessary? Does every opinion count? Or should it be majority rule? Or something else altogether?

That, too, needs to be debated and decided by whatever means can be agreed to, by whomever is involved. There won't be a perfect solution. But these are things we may decide on a case-by-case basis and apply to one-on-one interactions, relationships with friends, family, community, nation and so on. Not everyone will be happy. Not everyone may be included. But hasn't it always been thus? So, there is no point in pretending that we have actually reached the right answer, that the problem has been solved. All we can really say is that we have established the rules of engagement with such-and-such people, for the here and now.

This may not seem like a satisfactory answer, for it does not provide the degree of clarity or the level of certainty most of us prefer. What it does offer is a kind of freedom, however. Freedom from the restraints of perfection, freedom from the tyranny of illusions of moral clarity and absolute rules rather than general guidelines and practices informed by experience.

There is only us; and we must figure it out as best we can. We have to live with each other, navigate our lives in a way that benefits the people we care about, and if we are lucky, those we may only care about peripherally. It could include those that we consider mostly out of a sense of obligation or because we think we ought to care about them if we are good people; those whom we hardly know but understand may be suffering.

If we agree with the philosopher Patricia Churchland, morality is simply the byproduct of our sociality. And our sociality itself exists to keep us alive for another go at the genetic lottery. So, from that perspective we only owe each other what we decide to owe each other. And in practice, that often means we decide to owe our kin more than we owe strangers on the street, or starving children in far away places. Charity stays at home; everyone else gets our thoughts and prayers.

Is this what it means to be human?

20

We Belong to Us

Often, what we want most— more than fortune or fame—is to belong. We want to feel like we have a home, somewhere— not a physical one but a conceptual, social one, where everybody knows our name.

It is such an intense need that we feel lost when we don't belong, and we feel that loss acutely, as if nothing else will be able to fill the gaping hole in our being. We go so far as to imagine that others do not have this gap in their lives, that they have fulfilling relationships, particularly friendships that are close, warm, and reciprocal.

We wonder: Why don't I have that? Why are my friends not the same as other people's friends? What's wrong with them that they don't do for me what other people's friends do for them? And perhaps that's true, our friends are not like other people's friends. But it may also be a generalization— everyone's friends may very well be different, their relationships determined by their respective histories, circumstances and a host of other factors that are hard to pin down.

It is like a lot of things viewed from the outside. The inside looks cozy, rosy, idyllic— approaching a semblance of perfection. Or it is like looking down on a vast landscape from the commanding heights of an airplane window seat. Everything below appears seamless, flowing, harmonic. The faults are harder to see with the naked eye. Our perception is limited, our senses dulled into feeling like there is perfection, or something like it, far beneath us—perfection that doesn't seem to be there when we are on the ground.

So it is with friendships, or really any relationship that we are unhappy with. Other couples seem to be getting along famously when we are having trouble with our own partner. Other children seem better behaved than our own. This is, of course, an illusion, triggered in part by our circumstances; by our solipsistic, first-person perspective. Unavoidably, we know more about our own thoughts that about anyone else's, more about our own feelings than that of everyone else around us.

But we have theory of mind. We can think about what they are thinking, though we would be wrong, more often than not. Understanding other people is not easy. We make too many assumptions, based on what we are motivated to believe. We are unaware of the biases that lead us to believe some things and ignore others. It takes a tough-minded act of will to believe something different from what we have long believed.

So, we continue building the fiction, writing and re-writing a story of where we stand relative to others. And the grass continues to be greener on the other side. On the internet, it is even harder to shake the feeling that we are falling behind in all the ways that count. Other people seem happier, more glamorous, more successful; they appear to have more and better friends and relationships; go on more exotic vacations; have more exciting lives. Our own pales in comparison. We can't keep up. Perhaps we try, perhaps we even make a sustained effort but inevitably we fail. We fall behind, as we have always known we would.

And then we start sinking, going deeper into our own psyche and sometimes we can't find a clear way out of the darkness and back to the light. We can't come to terms with just being average, even if we understand the statistics, even if we can do the math and see that if our assumption is true—that is, if most people are doing things one way, having one kind of life—then by definition they are the ones who are average. And we are the exception, the ones that fall within the tails of a bell curve distribution of the population. We are among the few, perhaps unique in our own way.

But even if we sometimes believe this, it is cold comfort. We think we want what they have, or at least what they seem to have. Even if we read, even if people tell us, that it is all an illusion, that it is a performance for the audience, we can't reconcile it. We still feel envy, we are still afraid of missing out, of not belonging.

All of this points to a deep-seated fear of mediocrity, of being one of the masses. We have been told and have internalized the notion that we can do anything we put our minds to, that our reach will never exceed our grasp. We have bought into the great American dream— now the global dream.

Thus, we expect our life to be like other people's— not just ordinary other people, but like that of people we see on tv or in the movies; like that of the ones we read about or see on the internet. These are the ones living their best life with their best friends.

We forget the context of this information, we forget its purpose, its biases. We are reeled in by the storytelling. It is easy to fall for a narrative that we want to fall for. We want that life. We want their life— we imagine a level of fulfillment from their existence that ours has hitherto lacked. We anticipate a degree of happiness that we never thought humanly possible. We don't think of it as fiction or as a soap opera. It is reality through and through.

We understand at some level that we do belong. That at least we belong with our immediate family and perhaps even our extended family— if we are not estranged from them, that is. Yet, that isn't always enough.

Why? It may be partly a reflection of inherent needs, based on each person's personality and history, and partly a cultural phenomenon that leads us to desire what the culture at large values. Friends and friendship are a significant part of modern culture, so much so that beyond their actual value they are also performed for the benefit of a wider audience. That is, there is now a value in friendship that goes beyond the friendship itself because it confers social legitimacy and possibly higher status.

However, the conditions for friendships to form are varied. Nevertheless, it likely has a lot to do with opportunity to interact with someone, including casually and unexpectedly. The more times we encounter someone, the more likely we are to become a familiar face, to start a conversation, to find out more about them; and them about us.

But even then, the relationship may not set into anything resembling friendship. It could be an absence of shared affinities— perhaps there is a critical degree and number of affinities that need to be shared. Maybe it has to do with personalities, with some people getting along and bonding more easily than others. There could also be non-interpersonal factors, such as family or work commitments that don't allow for enough time or mental capacity to invest in a new friendship.

Perhaps most of our time is already committed such that we don't have the spare hours or the inclination to form, maintain and actively cultivate new relationships. Or perhaps it is none of things. It is nothing we can put our finger on. It is possibly another enduring mystery of human interaction— why some people make friends easily and others do not. Or perhaps it can all be reduced to pheromones and oxytocin and brain chemistry, which determines whom we bond with and whom we don't.

Despite our best intentions, perhaps we cannot help but dwell on our social status, on the state of our relationships. Maybe we cannot keep ourselves from comparing and contrasting, from feeling sorry for those who don't have what we have; or from feeling left out when we are the ones that others feel sorry for. When we are the ones that don't belong.

But studies on this topic suggest that the so-called circle of trust is quite small for most of us. It is often no more than five people, and of those, only a couple could be considered the most intimate, non-martial, platonic, adult relationships. Yet the messages we hear, the images we see, suggest otherwise and our ancient neural programming kicks in. The fear of being ostracized is so strong that we react at the smallest provocation, the slightest suggestion that we may not belong— because not belonging may once have meant the difference between life and death for our ancestors.

The thing is, relationships outside of family have a significant effect on both our physical and mental health. Having stable, strong social networks help us to weather the inevitable storms that are a part of life. They help us manage stress and help boost hormones that improve our sense of well-being while suppressing ones that raise our level of agitation. They determine the level of oxytocin, vasopressin, cortisol.

Friends and family also add meaning to our life. They help us feel as if our existence is purposeful, that we matter. Knowing that someone cares for us, will be there for us, especially in a time of need, settles our mind, frees us from at least some of our worries. It is a social safety net that is hard to replicate with anything else.

Yet, it is also important to realize that friendships naturally change over a lifetime. The thing that tied us with someone at some point in life, which created a shared affinity, may not be as big a part of our day-to-day anymore; and it may not loom as large in the other person's life, either. Thus, we may drift apart and as hard as we try, we may not be able to re-establish the connection that we used to have.

It is natural to question our identity when this happens. It is natural to keep coming back to ourselves, looking for reasons why there may be something wrong with us. To wonder what prevents us from making or maintaining relationships. But often it is more complicated. At some point all of us have to decide whether to keep trying to revive a lapsed relationship or invest in new ones, because there is an opportunity cost to pining for what has been lost. We may be better off opening ourselves to new social connections that can re-invigorate and sustain us rather than dwelling on the past.

21

On Our Own

Loneliness has long been a feature of human existence, sometimes by dint of choice, but often not. Either way, it is important to separate loneliness from social isolation— both are equally undesirable, but they are distinct. Sociologists define loneliness as the subjective feeling all of us have from time-to-time, and perhaps even for extended periods, when we are socially disconnected from others.

This can happen even when we are among people— both known and unknown. For one reason or another we may feel like we don't belong with the people we are with. Being subjective, loneliness is difficult to measure and any attempts to do so rely entirely on reports by the person experiencing it— we have no independent access to their consciousness to validate what they are saying.

For instance, we may not know a lot of people but if we have a handful of friends with whom we have a deep bond, that wouldn't be evidence of loneliness— even if we sometimes feel like those friends don't understand us. Social isolation on the other hand is objectively measurable because it entails living alone, reduced frequency of interactions and a lower number, and fewer types, of relationships.

Regardless of the difference, both loneliness and social isolation can have significant and lasting impacts on health, including increased risk of mortality. And by some measures, the risks of dying from loneliness and social isolation are higher than the risk of dying from a prevalent chronic disease such as diabetes. For instance, research by psychologist and neuroscientist Julianna Holt-Lunstead, suggests that social isolation can be as bad as smoking 15 cigarettes per day, in terms of increased risk of death.

Similarly, the 75-year longitudinal Harvard Study of Adult Development, concluded that loneliness is toxic to men in particular. The more isolated someone is, the more their cognitive function and physical well-being declines and the less happy they report feeling. In general, men are more likely to be living alone under the age of sixty and women are more likely to be alone after the age of sixty— primarily because of the difference in life expectancies between the sexes and the overall increase in life spans.

But it is not just the old that experience loneliness and social isolation. Overall, 28% of Canadians now live alone— the highest proportion in recorded history— and 10% of Canadians under the age of 60 also live alone. In the UK, the Office of National Statistics reports that 16- to 24-year-olds reported feeling lonelier than pensioners aged 65 to 74. There has even been a campaign to end loneliness and the British government at one point appointed a minister with responsibility for addressing loneliness. Psychologist, John Cacioppo, who pioneered what he called social neuroscience and research into loneliness, reported that up to 25% of Americans in the 2010s reported having no confidants at all.

Most people appear to miss simple things, like having someone to laugh with, someone to talk to or just sit down with or give a hug to. Older people report missing doing everyday activities with someone, like eating a meal together, holding hands, taking a walk, or going on holiday together.

As populations around the world— and especially in the so-called developed world— get grayer and grayer and as the rates of family estrangement and divorces grow, the risk of loneliness is only increasing. In fact, in places like Japan, which is known for having one of the world's oldest populations, lonely deaths among the old even has its own name— *kodokush*.

Those who study it say that loneliness has grown throughout the 20th century and into the 21st. Increasing prosperity has allowed adult children and newly married couples to live on their own; it has also allowed joint families to go their separate ways. Instead of having to fight over common resources, they can have dedicated access— no need to share. This nuclearization has re-defined notions of privacy, whereby we share less with those around us, even as we share more with strangers.

The content of what we share with the world though is often the version we want others to see— we hide away vulnerabilities and insecurities. We become addicted to being liked, which often means suppressing aspects of ourselves that we believe other people may disagree with.

We also trap ourselves in echo chambers, mostly interacting with those who agree with us and shunning those that don't. The new community is characterized by identity and ideology, it elides complexity, shuns dissent, shames readily. And it is in this environment that we seek and often fail to find meaningful social connection. Thus, we have only grown lonelier and lonelier in the decades since people started migrating from rural to urban environments and families started scattering.

In the not-too-distant past, most people rarely wandered more than a couple of hundred kilometers from the place where they were born.

But now, our so-called "life tracks" take us thousands of kilometers over a lifetime. We crisscross the continent we were born on and shift between continents more than once in a lifetime, sometimes willingly, in search of adventure (perhaps to fit in with the idea of how a person should be) and at other times unwillingly, driven by war, famine, or economic necessity or career opportunity.

Some even leave families behind to earn a living and are perhaps unable to return for years at a time while scraping out a meager existence far from their loved ones. For them, this is a preferable option, regardless of how they are treated in their new home or whether they feel like they belong.

Even as our life tracks take us further and further away from home, typically into bigger cities and communities, we are often left feeling more and more isolated. While some people successfully find and embed themselves in new places, others struggle. They become disconnected from more traditional forms of support, including extended family, religious groups, and cultural affiliations.

In the early years of the 21st century, the rates of reported loneliness grew— not just among the elderly but also among youth. Family sizes that started shrinking in the 1960s after family planning became more prevalent in the West, also led to nuclearization. Fewer and fewer family units now include grandparents or other relatives— they are primarily composed of the parents, children, and whatever pets they choose to have.

The increased remoteness of extended family relationships— such that children don't grow up as familiar with relatives outside the immediate family unit— combined with rising divorce rates and mixed families (composed of divorced partners and their respective children), adds to the sense of fragmentation and isolation. In fact, a subset of adults reports having no contact whatsoever with their families of origin— an estrangement that can be distressing and isolating, depending on the circumstances. Such drifting apart happens slowly and may not always be traceable to one event or happen for a clearly identifiable reason. However, it often follows periods of abuse, neglect, betrayal, or poor childhood experiences that may include a perception of parental betrayal.

We have been gradually but steadily turning more and more inwards— into ourselves and our own families. Everyone else becomes strangers, whom we interact with in mostly superficial ways. This may not be a problem if finding affordable, reliable childcare and elder care were not an issue; if people were not living longer; if both parents weren't working full-time, or if work itself weren't as all-consuming. At the same times, the families that we do have are shrinking in size, as fertility rates fall— in some places below the replacement rate (that is, the rate of births is lower than that of deaths).

All this is changing how we see ourselves as individuals, as well as our place in the family, community, and society in general. The trend towards focusing on individual vs. collective interests exacerbates our isolation and weakens the social glue that is one of the major predictors of long-term health and well-being.

Some studies suggest that the emphasis on individuals influences how we see the world, making us more likely to believe that our success is solely attributable to our own efforts, while other people's failures are entirely their own responsibility. Of course, we don't accept the reverse— that we could be responsible for our failures and others could contribute to our success.

It also seems to be cultural. Traditionally, eastern cultures have emphasized collective good and cooperation over individual benefit and going-it-alone. But that, too, may be changing as Western culture has gained a disproportionate influence on global culture— especially as worldwide incomes rise.

But some people are resisting such cultural forces. As different gender identities and sexual orientations have become more accepted, and as parents struggle to find childcare or elders face difficulties living on their own, new types of families are being formed. The very definition of family is changing. With these so-called "intentional families," people are choosing the members of their families and including people who are not related to them by either blood or marriage.

In some instances, they are coming together to buy homes and share day-to-day responsibilities, including ones related to domestic activities, childcare and elder care. These families of choice are new and few, but they may be an effective way to counter the lack of social connection and cohesion. Perhaps they will encourage thinking outside the self to consider how we affect other people. It may, in the most optimistic reading, also serve as a counterweight to the forces that are driving people into echo chambers where they become less and less willing to hear dissenting opinions.

Consider that all of us alive today are separated by, at most, 50 degrees— that is, we are at most each other's 50th cousins. We truly are a large global family. But like all families, the relationships therein are fraught. Often the hardest to manage are the ones that we have with those closest to us—parents, spouses, children, siblings— because we spend more time with them, because we know them better. These people elicit a full spectrum of feelings from us— feelings that we may be able or willing to control better with strangers— from intense love to intense dislike and anger.

It is, as yet, unclear what the future holds for family structures— whether we will continue to drift apart or start coming together in new and unexpected ways. Regardless of longer-term trends, if we accept the sociological definition, the self exists in relation to its social environment. That is, we are not fully ourselves unless we are among others.

While it is often said in half-jest that hell is other people, in some ways, so is heaven. If we believe that happiness is somehow worth pursuing, then based on all the available evidence, it would be very difficult to achieve without others— especially family members, however they are related or unrelated to us. It is, therefore, to our benefit to define ourselves more broadly as beings that incorporate our relationships and not just ones that are defined and confined by borders of skin and culture; by the physical / personal space we say we need; by the temporal rights— the "me" time— that we believe we deserve and must have. It is better to assume that we are not enclosed in impermeable Markov blankets.

Those suffering from psychiatric conditions are sometimes said to have difficulty separating self from the other, but sometimes there is value in ignoring that separation. Perhaps, we would all do well to make that distinction less concrete— to stop insisting on hard boundaries that define where we stop, and others begin.

22

Touched by Magic

As social animals, connections with other people seem vitally important to our well being, even if there are some who can handle and perhaps even prefer isolation. Here, too, we are not unique. There are other animals that rely almost as much on close interaction in a nurturing environment, which often includes close physical contact as well as other forms of non-verbal communication.

Experiments conducted by the psychologist Harry Harlow in the 1950s offer a glimpse into the significance of these factors. Working with monkeys, Harlow separated newborns from their mothers and assigned them to surrogate "mothers" made of either a wire frame or a soft terry cloth. Depending on the experiment, both surrogates could dispense milk to the baby monkeys through an attached bottle.

However, when the surrogates dispensed milk and the babies had a choice, they tended to prefer the terry cloth "mother." But even when only the wireframe mother provided milk, the babies would drink from "her" and then go cuddle wit the terry cloth mother.

When the babies were only assigned to one or the other surrogate, and each provided the necessary nutrition, there was no discernible difference in the physical growth and development of either set. However, there were significant differences in their behaviour.

For instance, when they were frightened by loud noises that the experimenters made, the monkeys "raised" by the terry cloth mother sought comfort there, rubbing themselves against her, soothing themselves and gradually calming down. The ones raised by the wire frame mothers, on the other hand, did not seek refuge in their "mother" when frightened. Instead, they clutched themselves, rocked back and forth and screamed, apparently incapable of soothing themselves.

This suggests that at least for babies and young children, physical contact with their caregivers is important to their ability to cope with the world, to self-soothe, become resilient and develop the confidence necessary to handle unexpected situations.

Indeed, touch is the first sense we develop, starting in the womb and as early as the embryonic stage, even though it would appear that there isn't much opportunity to exercise it in the womb. From there, touch continues to be a natural and important aspect of interaction among children and, increasingly to a lesser extent, among adults. However, some of this is cultural as there are both high-touch and low-touch cultures.

Already social concerns about inappropriate touch and harassment have discouraged all kinds of touching and experiences of local and global pandemics in recent times may further move all cultures away from practices that involve touch, including as part of greetings and goodbyes.

But clearly, people who lack regular social connections with close family members and friends are feeling the lack. This has led to the emergence of personal services that offer cuddles and hugs— for a fee. It is unclear if this will be sufficient replacement for regular, emotionally meaningful contact; or how it will affect the well-being of both those who seek and those who offer the services.

And this isn't just about touch as a sexual experience. In fact, it may be more important to experience regular touch in non-sexual contexts, particularly for those that have grown up in environments that discourage or frown upon it. Touch can clearly be one— but not the only— antidote to loneliness and social isolation. It is physical evidence of a connection between people, especially when it is non-violent and invested with emotion and meaning. It can express a kind of solicitousness, a very powerful gesture of care.

Touch has been shown to have all kinds of benefits, including in regulating hormones like cortisol and oxytocin, levels of which change in the presence of stress. By depressing cortisol, touch helps to lower blood pressure and heart rate. Increased oxytocin levels help to manage fear and have a positive influence on feelings of trust and generosity.

As a sensory input, touch is processed much faster and some of the effects can be felt much sooner than say hearing or sight, which go through much more complex processing and interpretation in the brain. Thus, there is often an immediate response mediated by the autonomic nervous system— it is the calm that we feel when someone holds us after a fright, it is the easing of tension in our muscles when someone shows affection through a gentle touch or stroke of the arm or face.

The nerves that detect touch are also more complex than was once thought. Touch is ultimately a mechanical stimulus and involves detection of pressure, texture, vibration and other forces and qualities. Different nerves are involved in detecting each of these. One particularly notable nerve ending, *c tactile afferents*, specializes in detecting the sensation of being stroked— a sensation that is almost universally perceived as being pleasant and desirable. Animals like monkeys and apes engage in it as well and grooming and licking can perhaps be seen as forms of stroking.

Whether as babies or adults, we generally respond to stroking and other forms of physical contact with positive feelings. And even contact from strangers, unless it is unwanted, can trigger such feelings in us. Something as simple as a light touch of the elbow or a tap on the shoulder makes us feel a connection. It keeps us well and alive. It makes us feel like we share something human; makes us feel that we belong.

23

(Be)longing to Work

Since the wide-spread industrialization of the 20th century, work and self-worth are inextricably linked for many people— arguably in the West and mainly for those in so-called white-collar jobs, but perhaps for selected others as well. Even as feminism has helped to promote women's rights and opportunities, it has also led women— like men— to associate more and more of their identity with their work.

Work has ceased to be simply a job— a way to make a living, to make enough to feed the family; to save for retirement. It is something that is expected to be meaningful, fulfilling, filled with passion. But beyond that, it has also become a necessity, that without which it is impossible to survive in the modern world— at least to survive in any practical way, without the generosity of others or inherited wealth.

Economics has ascended to primacy. If we do not work, we cannot easily get food, water, energy, shelter— the basics of human survival. For most of human existence, life used to revolve around finding, making, and consuming food. Then, we moved onto looking after domesticated animals and tending to crops, which we cultivated, harvested and processed ourselves.

Over time, trade and economic activity expanded. And we had more than we needed for our families and kin. We built cities and created civilizations, with all the complexity that entails. This is not to gloss over the tremendous amount of change that happened over tens to hundreds of thousands of years. It is mainly to point out that work as a distinct activity is a relatively new idea, and its association with self-worth perhaps newer still— definitely the extent to which it manifests today.

Even 300 years ago, most people worked close to where they lived and while the feudal system in some parts of the world meant that they had to pay to rent land, they largely worked for themselves. This began to change in the 18th century, when work started to become more centralized. That is, workers were gathered and worked side-by-side in one location. This is sometimes called the "industrious" (as compared to the later "industrial") revolution.

The 19th century brought the industrial revolution, initially to Western economies. Automation became prevalent, allowing work to be scaled much more quickly, efficiency to be increased and output to be greatly expanded. Paying workers a standard wage, rather than a share of profits, became much more common.

The two world wars no doubt changed how work was viewed, incorporating patriotism into both the need for, and purpose of, work. In many countries, with men away on the front lines, more and more women were recruited into the workforce. Work became a source of pride, of nationalism— a means to contribute to the war effort and to national glory.

After the second world war and the great depression, the West entered a period of stable employment. Many people found jobs for life— working at the same place for most or all their careers. This engendered employer loyalty, which that was often reciprocated, with many large private companies taking care of their employees into retirement.

The cultural upheaval of the sixties, combined with the cold war, initiated the end of the so-called "secure wage" from the 1970s onwards. Stable employment was no longer a feature of many people's careers. Services, and especially financial services, became a bigger share of the economy compared to manufacturing. And financial metrics became the dominant measure of a company's success, not just in the long-term but on a quarter-by-quarter basis.

All this created incentives that may have ultimately become harmful to our mental and physical health. Aggressive management to meet financial targets, the rise of market analysts and earnings guidance and the myth of the superstar CEO and worker are all trends that have emphasized work over practically every other activity in life. Or at least posited work as the means to achieve or pursue anything worth pursuing.

And as outsourcing and restructuring became fashionable through the 1980s and 1990s, companies no longer offered de facto lifetime employment. However, they still expected employees to remain loyal. In some industries, and increasingly with the rise of the so-called gig economy, work has become even more precarious. Employees can no longer expect predictable work or income. They are expected to be available on short notice so that companies can be flexible and nimble to market changes while minimizing costs.

The ready availability of technology to handle the logistics and management of such a contingent workforce has allowed more and more organizations to adopt such models, including in manual labour, education, and retail.

By some estimates most new jobs in the second decade of the 21st century were of a temporary nature with a third of all— and over half of young— workers engaged in precarious employment. For some, this is a positive development, especially if they are in high-paying professions, have other sources of wealth or are young, healthy and not averse to risk.

Temporary employment allows them the freedom to try many different things rather than settling into a long career at one organization or in a specific field. They can be their own boss, or even if they must report to someone else, they can easily switch jobs if they don't like their managers or colleagues.

For almost everyone else, precarious work puts them at an even greater economic disadvantage. It reinforces existing inequalities and makes it harder and harder for them to climb the socio-economic ladder to some semblance of stability and prosperity.

Work continues to be a big part of our lives. We are in an age of careerism and credentialism. Those of us that have the time to navel gaze and the luxury to ponder alternatives can hardly separate our work from our life outside of work; our identity from our supposed professional achievements. We are loath to admit that these accomplishments are hollow, that in the long run they will leave us feeling empty. They are like refined carbohydrates that offer a sugar rush but soon leave us hungry and wanting more.

Work has its uses, however— we learn things, we may meet friends, spouses, future business partners there; we may even change something for the better. But often, we don't change the world. We play our role, we roll forward; we turn the cog a fraction, but that cog is among many other cogs within a larger mechanism of gears and switches. We may not even understand what our single turn of the cog is for. We may not know or ever find out what our purpose was in the overall mechanism. And even if we did, it may not be all that significant. And that, to many, is ultimately disheartening. It is disillusionment.

To compensate, we end up concocting stories about the importance of what we do, about the impact we are having. Even as corporate communication serves as a kind of propaganda— both to its customers (marketing) and to its employees (employee "engagement")— employees create their own propaganda. They are trained to think that people should become products and market the product effectively to find success at work. They strive to brand themselves, to project a certain image of success.

The more ironic part of this advice is the focus on being authentic and humble. Authentic leaders, authentic followers; humble servants to your customers. Of course, this fad encourages everyone to prove their authenticity and humility credentials— and often the way they do it comes across as neither authentic nor humble. It cheapens both the words and the ideas they represent.

There is no better or faster way to lose credibility in projecting certain characteristics than to insist that we have them and that we are exhibiting them. There is no better way to demonstrate insincerity than by providing a constant stream of proof. We are hardly true to ourselves when we act this way. We hide our real feelings; we refrain from saying what we are really thinking.

By necessity, all of us adapt to our environment. While some personality traits may be stable over a lifetime, situations do seem to influence a lot of our day-to-day behaviour. As such, it is not surprising that we would behave one way at work and another way at home or with close family and friends. As Virginia Woolf wrote, "different people draw different words from [us]."

It is natural to vary our behaviours from one environment or situation to another. What this means is that it may not even make sense to talk about an authentic self if there is more than one version of us—one that makes the most sense for a given setting. And to invest so much time and energy in projecting an artificial image at work to get ahead in a career may be both disingenuous and bound to fail— granted some proportion of people who try it will indeed succeed.

Projecting an image— of authenticity or some other seemingly desirable characteristic— can sometimes be a stretch, since it takes focus and effort. For instance, we may be naturally animated, expressive and outgoing with close friends but wouldn't be so in a work environment where our relationships are mostly practical and superficial.

For some it won't be difficult but for others it will require energy that leaves them feeling drained. In the latter case, we should question if the effort is worthwhile; decide whether we should continue to cater to what we think people want from us. Or if we should behave in ways that are more natural— within acceptable social norms for whichever culture we are operating in.

We invest too much of ourselves in work. It is no longer just a means to make money to live on. It has become a reason for being. The result is that when things don't go well at work, when we are not recognized to the extent that we feel we deserve; when we are not rewarded or valued, we feel like we have failed. Then we start to believe that failing at work means that we have failed in life.

As much as we do and have to work, it is only a slice of our lives, and if we allow it to become too big a slice, we will shortchange not only ourselves but our families, too. When employer loyalty and beneficence is not a given, when employers only pretend— even to themselves— that the workplace is a family, we cannot afford to emotionally invest in them, to put all our energies there in the hopes of being treated the way we would be treated among our real family and friends.

There is often far less room for failure, for missteps at work than there is at home. Ultimately the organism— the company— takes precedence over the individual. Every employee, including the leaders and perhaps even the owners, are dispensable. Which is why it is so striking when we prioritize the existence of the corporation; when we are so willing to harm ourselves for its continued survival and prosperity; when a mere fraction of the returns of whatever it earns flows back to each of us.

Even if we do not dedicate ourselves to one organization, we dedicate ourselves to the profession— and while in some cases this is worthwhile and necessary— it need not be all consuming. Because if it is, we may exhaust ourselves to the point of not being able to meaningfully contribute to our profession.

We need to let go of the idea that work is the only meaningful way to live a good life. We act as if a life is not well lived unless it is recognized as such by the social standards of the day, which are often fickle and change over time, and which may, in any event, disproportionately value superficial achievements and ostentatious indicators of success.

We also need to think about how we rear children. Perhaps it has long been thus, but we seem, in modern times, to be living vicariously through our children. They are effectively extensions of our identity; they reflect our own sense of self and therefore— we believe— determine how others perceive us.

Consequently, we spend a lot of time trying to turn them into our notion of a perfect human being, which now means turning them into marketable products that can generate the highest returns, partly through work, and partly through status (which may in turn be dependent on their work).

This may entail ensuring that they wear the right clothes, go to the right schools, associate with certain kinds of people, develop networking skills, engage in appropriate activities, and so on and so on. We also try to prepare and steer them towards careers that will both reflect well on us and set-up them up for the kind of success that we believe is necessary to matter in the world. We want them, and therefore ourselves, to be seen as worth something.

Do we do this more for ourselves than for our children— to keep up with the global Joneses? Would those resources— including both money and time— be better allocated to something more meaningful. Fame and success— whether ours or our children's— is doubtful and fleeting. And is making children into products the best way to prepare them for the future?

It would seem so because there are many aspects of modern life that are fraught with risk, or at least perceived risk. Perhaps none more so that rearing a child. There is of course the self-evident risk to the child's safety, which by many accounts is vastly overblown, but there are also more subtle fears that parents feel and face.

Among these is the fear that one's child will fall behind, be less successful in the world than other people's children. The unquestioned Western dream is based on a belief in unlimited possibility, where every child can be an academic genius, athlete, entrepreneur, and high-powered socialite—if only . . ..

Or at least parents feel that it is their job to make their children into such multi-talented humans. This feeling is likely rooted in peer pressure (everyone else is doing it) but may also be entangled with ambition and a need to compete. The ambition is partly vicarious, perhaps an attempt to compensate for perceived shortcomings in our own lives. So, we push our children, over-schedule them, in the hope that they can be more than we were—because all of us feel inadequate in some way, as if we have somehow failed to achieve our full potential in life.

The value of all this striving is questionable. By all accounts few of the most accomplished scientists, artists and athletes in the world had what we are all trying to give our children today—cross-training in a multitude of fields; unparalleled exposure to the world's sights; freedom from boredom; intensive training in multiple sports.

Are we truly creating better humans or simply more exhausted ones, who can't think for themselves, whose every move is scheduled and who may eventually lack the motivation to do much of anything at all? Are we rearing children, who, perhaps most importantly lack the freedom *of* boredom rather than freedom *from* boredom?

We refuse to consider that perhaps children are their own best teachers and that our role may be to merely nudge them from time to time, to expose them to new experiences and ideas without exhausting them. Or perhaps our role is to do nothing of the sort, to only feed them and take care of them, and let them be—to discover the world for themselves.

While it may not necessarily be a formula for nurturing genius, it certainly doesn't seem to have hurt the likes of Newton and Einstein, who were by all reports prone to wandering (and perhaps "wondering") ways. Besides, all the generations of humanity before now didn't benefit from hyper-attentive parenting and yet here we are in the Anthropocene, with the power to both create and destroy worlds.

But the future is even more uncertain, and it is doubtful that the kinds of abilities and skills that have been seen as keys to success in the past will be as useful in the future. The accelerating pace of technological change and, in particular, the development of automation and artificial intelligence, is likely, within decades— perhaps in our own lifetimes— to change the nature of work in fundamental and unexpected ways.

In fact, it is quite possible, that much of what we call work today— the administrative make-work of most so-called professionals— will become unnecessary. And even the readily classifiable work of extracting resources and making things could be largely automated. What will we think of ourselves then— especially those of us that tend to engage in navel gazing?

Unless we are the inventors, maintainers or owners of these new machines and technologies, and the enterprises that use them, how will we make a living; and will we even need to? What if economic models are upended and labour— or at least human labour— becomes an insignificant factor of production?

We will have to look elsewhere for how to define ourselves. We can no longer rely on our work to tell us who we are and what we are worth— at least not many of us. The crutches will have been kicked out from under us and we will teeter on the verge of falling unless we can find meaning elsewhere, where it has always existed and been ignored. We will need to go beneath the surface of how we measure the value of a human being. We will need to rethink whom we admire, whom we wish to emulate and why. We will have to be inevitably, and more wholly, ourselves— and all that entails.

But then, these predictions have been made before, perhaps most famously by economist Maynard Keynes in 1930, who thought that his grandchildren would only work 15 hours a day and wouldn't know what to do with all the leisure they had. So perhaps it is not so much that work itself will disappear but that its definition will change— we will occupy ourselves in different ways, we will segment time into categories we can't even imagine today.

24

We Are Here

We are in the universe, in this galaxy, in this solar system, on this planet, this pale blue dot. We are spinning, revolving, moving through spacetime at incredible speeds but are none the wiser. We may as well be oblivious to all of it, as we were for thousands and thousands of years.

As the astronomer, Carl Sagan, put it, everything that ever happened to us, in the history of our species, happened on this pale blue dot. He said it after looking at an image of the Earth taken by the spacecraft, Voyager 1, from a vantage point beyond Neptune, on February 14th, 1990. At that point, it hadn't even left the solar system and yet the Earth was an almost indiscernible dot in the vastness of our small galactic neighbourhood. From even further out, it would be practically invisible without a very powerful telescope. A distant observer would only be able to infer its existence from other data such a dip in the light emitted by our sun as the Earth transits across it.

It is perhaps easy to say that in the grand scheme of things, nothing we say or do in life matters, if all we are is bits of matter, temporarily glued together. But day to day, it is hard not to feel as if everything matters, that we (individually and collectively as a species) do important things. From what we eat, how much we exercise and how we feel, to our politics, economics, history, and philosophy.

It is easy to get carried away with the tide of self-importance when everyone around us reinforces it, when we write tomes, discuss and debate endlessly about our policies, decisions, opinions, rights, freedoms. A lot of these are ingenious if abstract concepts of our own making, including our ethics and morals, our principles of justice, our philosophies. We place great stock in our reason, in our success in understanding and harnessing nature to our own ends. But it feels necessary, vital even.

Is this just human exceptionalism? Why should we matter? And if we don't matter, why should anything we do matter? If what awaits us is insignificance and oblivion, why do we fight so fiercely? Why do we kill? Why do we lust after fame and fortune?

Even on our own planet, most of us—in fact the vast majority of us—are destined to toil away in obscurity during our lifetimes. And if we achieve— even the word, "achieve," connotes importance, significance, worth— something in our lifetime, most of us will soon be forgotten, end up in the recycling bin of history. If we are lucky, we may be dredged up by some obscure future scholar looking for a lifeline or by a curious soul sorting through the trash.

But even then, our revival may be short-lived. In the very long run, we are, almost all of us, destined to be forgotten. In a mere hundred years from today, most of us alive will be dead; there will be all new people. So why do we go through all the trouble when we are alive?

Perhaps because at some level, we need to feel as if we matter. Because if we don't, we would sink into an abyss of despair at the realization of our irrelevance, at the knowledge that in the final analysis it likely would not have mattered whether we had ever existed.

It may be useful to keep that knowledge in the back of our minds and death by our side. Because it is that irrelevance that can free us if we embrace it. It can help us to find a measure of freedom during our brief, blip of an existence in spacetime.

It is useful when so much about who and what we are, and even what the world is, remains unknown and uncertain. When even the fundamental nature of reality and of our conscious experience of it remains unresolved, it would be easy to lose our minds, to not know left from right, as it were, if we dwelt too long on the nature of the universe.

It is one thing to contemplate all these things in abstract, philosophical, or mathematical terms and quite another to face the full existential force of that knowledge and to feel the complete dread of it. It would be difficult to remain sane for very long.

Thus, there is, on a practical level, the need to matter, to be important, or at least to act as if all of it were true. This is the concept of dharma, of doing our duty, which, although it can be interpreted in many ways can be thought of us "that which we must do." But perhaps we can do our duty in full knowledge that it may be meaningless, that it may ultimately serve no purpose.

We are naturally uncomfortable with purposelessness, with meaninglessness. We have spun such stories about ourselves, ascribed so much meaning to our existence and to everything we do that any hint of futility triggers a strong reaction. We yearn for ultimate purpose, invent, and then invest in complex rituals and religions to make it seem true. But all of it may be nothing more than sense-making, the story-telling that we engage in every day.

And that, too, may be something we need to do. Or perhaps something we are compelled to do, no matter our efforts to stop. This points to another aspect of our existence that is closer to home—our shared history and biology with other living things. We would do well to remember this every time we feel the urge to assert our dominance, to insist on our exceptionalism and use it to justify all manner of harm that inflicted on our fellow space travelers and poison the well from which we drink.

We should recognize that we are part of an ecosystem and that entails being part of a food chain, whereby we must kill other living things to feed ourselves, to harvest the energy that they have obtained and stored. For all our ingenuity, for all the innovations that allow us to obtain power to build run machines, we are as yet unable to use that energy directly, to power our bodies. So, until we find that direct pipeline between non-living sources of energy and our own bodies, we have to get it by killing and eating plants and animals.

While we can acknowledge and act on this necessity, we also now have the freedom and capacity to do it in a way that minimizes harm, to avoid unnecessary suffering. For all our arguments about consciousness, we have no way of accessing the direct experience of another human being, let alone a plant or animal. And in the absence of that direct knowledge, we lose nothing by acting as if other living things are indeed conscious, to whatever degree. That is, we could act with a measure of compassion towards one and all— human or otherwise.

Often, what gets in the way of compassion is all the abstract concepts we have invented, which help us wrap ourselves in webs of significance. All that stuff about our individual rights, even insisting at times that such rights are exclusively granted to our species by some imagined, omnipotent overlord.

That brings us to the question of reality itself. If quantum mechanics and quantum field theory hold, then reality is ultimately a probability distribution and is much, much stranger than we can ever imagine. Whether it is one universe or many, one world or many worlds, we can neither see nor know all of it. We are like the smallest microorganisms limited by our primitive senses to only ever perceive a fraction of everything that's out there— whatever out there means or is. In our own way, we may be very limited, as limited as many of the animals—

and sometimes people— that we look down upon, disregard, destroy.

All this should make us humble. It should make us feel small and insignificant, or no more significant than anything in the observable universe. We are not ascendant. We are only here, for a time. And our time will inevitably run out. Our species will not last forever, regardless of how much we discover, no matter how much we invent. It seems highly unlikely that we can innovate our way out of our existential fate— to live out our few million years as a species and then give way to whatever comes next, if it is anything at all.

Either way, the energy or matter of which we are composed will dissipate, perhaps reconstitute as something else, or perhaps disperse slowly, inevitably into the unimaginable vastness of spacetime. Perhaps it will re-emerge in another universe, or simply, eventually be extinguished, to be replaced with nothing.

III

Becoming

25

Becoming Who We Are

In going from who we are— or think we are— today to who we want to become, it helps to understand that we are not alone. That is, we are not an island in a vast ocean. We exist in a universe that is not empty save for us. We exist in nature, in an ecosystem of living and non-living things; we exist in society, among other people. Everything we are arises from this context; it does not emerge from a vacuum. Thus, there is little value in over-individualizing our behaviour, desires, thoughts, or feelings. We are, in essence, a response to our environment. And that is true for every living thing. We wouldn't be us, without us being in the world.

Our fears and hopes, our neuroses and our genius, our successes and failures are all part of our response to the world and intricately and inextricably connected to it. We are not solely responsible for these things. We can neither take all the blame nor claim all the credit. Our life is an exchange of energy and information with everything around us. And perhaps energy and information are the same thing. We are not an independent entity with independent thoughts; we are influenced by people and situations, genes, and cosmic constants. The causal chain of our thoughts and actions go back further that we can glean to the very origins of spacetime. The beginning of everything is always much earlier that anyone thinks.

In looking forward, we need to take that into account. We need to consider our environment— the places, living things, people, and situations that we may encounter. We need to think about the things that could affect what and how we think, what and how we feel. Who we are, even. Our plans need to consider this context, leaving aside for the moment whether our plans are our own or if we even have the free will to make plans.

We also need to account for randomness. For the fates, as it were; or more accurately, we need to consider the role of entropy and probability. Much of our ability to realize anything at all will depend on pure, dumb luck, although it is neither pure nor dumb.

Because society holds an individual responsible and accountable for their actions, it also makes them solely responsible for their failures. Thus, our shortcomings are seen as our fault and we will tend to blame ourselves, deny self-compassion or fail to see the role that systems and people in our environment may have played in any given outcome. But the reverse is also true. We influence others; we are partly responsible for how they behave. We are enmeshed.

On the other hand, there is success, which is frequently the goal of becoming —no one strives to become a failure. We imagine a trajectory from where we are now to somewhere else where we are better, on whatever dimension that appeals to us. To become better is to have succeeded in our striving. And we often attribute that success to our own choices and actions. It is an idea as much as an achievement. But it is amorphous— it means different things to different people. Nevertheless, it has one thing in common— recognition and adulation by others. Success isn't success if other people don't know about it.

Consequently, much of our adult lives now seem to be about running a race to reach an imagined podium of our dreams. We think that this race can be won; that it must be won, without necessarily grasping what the race is for. It is as if we were suddenly dropped onto a treadmill and had no choice but to run to keep from falling off.

The race isn't simply about getting into the best school or climbing a corporate ladder, it is also about having the perfect life, the best trained children, the twice-a-year vacation, the right hobbies, the perfect habits, the best stuff. It is about being able to have those meaningless hallway conversations and knowing enough to laugh at unfunny jokes, so we feel like we fit in.

We believe that these are all necessary for staying in the race, to keep up with the world full of Jones's, to avoid falling into irrelevance.

But is that so? Is there really a better us to be had; a more perfect us? Are we in the process of becoming something more than we already are? Are we on a journey to the ultimate? If so, consider that perfection, by definition, is unattainable. Once it is achieved, it ceases to be perfect. So, the race is a Sisyphean one, if ever there was one. And we must decide whether we want to stay in it. And, if we do, then we should ask why? If it is for bragging rights, then can we really claim bragging rights when everyone else can also claim them?

All this is not to say that striving is worthless, that we cannot become more than we are, today. Only to pose the question of whether chasing someone else's idea of perfection is the best use of our time. Instead, consider striving as an end in itself. Why not do things for their own sake—not simply because of how they will make us look or feel in relation to someone else. After all, we don't know what the future holds or whether the race will be worth it in the end.

Life is most navigable in small increments. Our expectations about the world, our ability to predict anything at all has the greatest chance of success—though it remains abysmally small—if confined to the next few minutes.

What we may become is not entirely within our control and what it means to be us has already started to change. Some philosophers suggest that we have already extended our consciousness outside ourselves, for instance by using books, art, and computers to record our memories and thinking and using such things to navigate the world in ways we never could before. Are those things also a part of us? Does how I define "me" include the things I have created or the tools that I use, be they hammers and wrenches or books and computers?

Already we are extending that metaphor even further. We can use prostheses, with sophisticated ones being able to receive and process nerve signals to move an arm or leg. Cochlear and retinal implants can help people hear or see at higher and higher resolution and fidelity than ever before. And at the most basic level, technologies like CRISPR allow us to edit specific genes in an embryo, to potentially choose some characteristics of the person that is ultimately born— characteristics that were not present in the baby's parents.

What will we have become at the end of this road, if scientific discovery advances to the point where the specific neural connectivity, structures and properties within an individual brain could be mapped and re-created, preserving whatever memories it has accumulated, as well as whatever capabilities it has developed? Could we then be re-created as a machine or perhaps a cyborg with a machine brain and a human body? Would that still be us? Will we finally have arrived at our destination? Will we then have realized our destiny?

Perhaps. But it helps to remember that we never stop changing, that we never stop becoming something different than what we are now. This happens because our environment is always changing and we, being its creature, are constantly adapting—until, that is, we run out of road.

And the changes we experience aren't always teleological—despite our most ardent hopes, change doesn't always equate better; it doesn't necessarily mean progress. That is also something we need to reconcile ourselves to, the idea that self-improvement projects don't have to lead to improvements, partly because the universe is random and partly because we are inconstant.

26

Letting Go

Buddhist philosophy suggests that we suffer because of our inability to let go. We hold onto notions of control, we covet material possessions, we desire happiness and pleasure—and it is that desire that makes us unhappy.

But impermanence is the natural order of the universe— nothing lasts forever. If we hold on, staying attached in a world where things are fleeting— whether it is fame or fortune, successes or failure, and even life itself— we are creating the conditions for continued suffering.

Whether or not we subscribe to this particular view of existence, it has useful lessons for day-to-day living. And practicing detachment doesn't need to go as far as giving up all material possessions or living as an ascetic in a cabin in the woods or atop a mountain.

The Tibetan word for attachment is *shenpa*, although it could also be translated as "hooked" or an "urge." The Buddhist teacher and nun, Pema Chodron calls it the "sticky feeling." This attachment relates to more than material things— it also has to do with feelings and ideas, about us and about others.

The *shenpa* rears its head when someone does or says a thing we don't like. We don't always see it coming. It arises suddenly, unbidden, and it grabs on, refusing to let go. The feeling is difficult to describe—a combination of irritation bordering on anger, indignation mixed with resentment, shame compounded by guilt. We feel compelled to respond, to strike out, to punch back.

We don't want to be called out this way, we don't want to be revealed, we don't want to expose our weaknesses. We don't want our secret insecurities to be laid bare for the world to see.

It feels personal. As if they are targeting us specifically, not generally. They are out to get us. This feeling fuels our indignation. It stokes our irritation into full-blown anger. How dare they, we think. Who do they think they are? These are not sophisticated feelings. This is not the Cartesian theatre, it's Shakespearean.

Arguably, material attachment is easier to shake than this kind of psychological attachment. Feelings of anger, jealousy and hurt are harder to let go. There is perverse satisfaction in holding on to them, to nursing and revisiting them, to nurturing and feeding them. Resentment feeds resentment.

However, such reactions can become habitual and thus even harder to get rid off— we get hooked. The response to a particular stimulus becomes quick, automatic System 1 behaviour rather than a slower, more deliberate System 2 one.

Often, we are most attached to practicing these habits on those closest to us. They are the ones who trigger us the most, the ones whose habits we are intimately familiar with and sometimes have the lowest tolerance for. We get hooked into and invested in our reactions.

When the *shenpa* gets is claws into us, it doesn't let go— it is hard to get unhooked, even when we realize it is happening. It happens even when we can see ourselves falling into the hole, giving into the habitual reaction—the momentum is too great. To paraphrase Newton's third law, a feeling in motion stays in motion unless stopped by an equal and opposite feeling.

But in the long run, we rarely feel good about ourselves for indulging the urge to react as we do, for remaining attached to those negative feelings. Through it all, we may rationalize, go to great lengths to justify our own actions, even as we condemn those of others.

It may, over time, have even become a means of self-soothing, of feeling good about ourselves, especially in times when we may not be seeing ourselves in a very positive light.

Thus, part of the *shenpa* is judging and comparing, telling ourselves that we are better— more moral, disciplined. But all that does is help us hold on, stay hooked, attached to those not-so-helpful feelings.

Yet, getting rid of it, moving past the urge to respond as we have trained ourselves to respond, is easier said than done. But it is not impossible. We could start by simply recognizing that it is happening, that we are getting hooked and acknowledging that it is a habitual response, which may not be entirely justified.

Once we have learned to recognize that it is happening, the next step is learning to pause, to simply interrupt it. For some people, this is where meditation and mindfulness may be helpful, but admittedly, that is not for everyone.

So even if we are the kind of person who doesn't believe in or can't get into a formal and regular meditation practice, we can still teach ourselves to take a breath and simply pause. In meditation, pausing involves accepting what is happening and sitting with the discomfort of it, letting it pass over us rather than yielding to the desire to eliminate the unwanted feeling.

It may be helpful to pre-determine a specific something we will do when we recognize the *shenpa*. It could be leaving the room, saying out loud that we need to take a moment. It could involve doing something silly, fun, or relaxing. We could also write down our thoughts or tell a neutral party. The key is to do something that we enjoy, to distract ourselves, to release the pressure that's built up, to take our mind off the trigger and give ourselves time to let go.

But pausing can also be as simple as breathing— not necessarily breathing as we would during meditation, but taking deep, slow breaths. It turns out that for something that is essential to our survival, most of us don't breathe in an efficient or effective manner. Our breaths are too short and our breathing too rapid— we are shallow breathers. The serious implication of this is that it prevents us from optimally balancing not only the oxygen, but also the carbon dioxide in our blood.

Instead, we can learn to do what's called "belly breathing," also known as diaphragmatic or abdominal breathing. This involves a slow intake of breath through the nostrils, with the mouth closed, using the muscles in the abdomen— rather than the ones in the chest— to draw the breath in. Then, exhaling through the mouth, lips slightly opened; this time relaxing the chest and expelling the air by contracting the abdominal muscles again.

We can time each breath— in and out. Breathe in for four or five seconds and breathe out for the same duration. It may also help to put one hand on the belly and the other on the chest. As we breathe in an out, the hand on the chest should remain still— only the other one should rise and fall.

Learning to breathe properly has a wide range of proven physiological benefits— from lower blood pressure and heart rate to an enhanced ability to mitigate— and in some cases eliminate— conditions such as asthma, chronic obstructive pulmonary disease (COPD) and allergies. Psychological benefits have also been noted, including improved ability to deal with anxiety.

Additionally, deep breathing has been shown to calm the brain's arousal centres in mice—and likely in humans as well. In some monks who have extensive experience with it, deep meditation has been shown to suppress the startle reflex. Paying too much attention to our thoughts activates the anterior insula and the anterior dorsal cingulate, an area of the brain that is connected to motivation and anxiety— also called the salience network (SN).

But the SN is also what allows us to determine the relevance of stimuli (which would otherwise overwhelm us if taken in all at once) and coordinates the response to such stimulus. For instance, it is tuned to amplify new and unexpected information, or information that we have learned (through evolution or training) to prioritize.

Disruption of the SN may partly explain diseases such as autism (where social information is not prioritized) and Williams disease, which results in hyper-attention to social information. It may also play a role in motivation, as rodents whose salience network is damaged give up more easily when trying to find food.

The SN, in turn, is connected to the so-called "Default Mode Network," (DMN) which is activated when we are reflecting, focusing internally, reminiscing or daydreaming. There is some evidence that the DMN prepares us to act even though it only becomes engaged when we are not actively thinking or planning.

Jut as a meditation and breathing helps us learn to pause and become aware of our thoughts, a body scan develops awareness of our body. Going from head to toe, we can learn to focus our senses inward, progressively noticing each part of our body and consciously relaxing any tense muscles.

Then, we can send our attention outwards, engaging the senses, one at a time. We can start by feeling the floor we are standing on, or the surface of the seat we are sitting on. We can listen to the sounds around us; and take in the space with our eyes.

As with meditation, we should simply observe, not engage, process, or judge. It will help us get ourselves in tune with our body and our senses—inside and out—and prepare us to engage the world in a more present, composed, and meaningful way.

Beyond such new age seeming techniques (although many are

hundreds and thousands of years old), there is always traditional prayer. Surprisingly, prayer has many similarities to meditation.

Indeed, one way to still the mind is to engage it frequently in a repetitive activity—for example, through a chant, rocking body movements or counting beads. Ironically, engaging the brain in this way also focuses and helps to make it still. Meditation just does it in a more powerful way that has proven benefits—including, changing the length of our cells' telomeres, which play a major role in cell division, and which seem to be related to lifespan (telomeres shorten as we age and approach death).

However, we do it—through breathing, a body scan, praying or something else—it is the pause that matters— not how we pause. This is so at least from the narrow perspective of learning to detach ourselves, of learning how to create a space between stimulus and response, which the psychiatrist, author and holocaust survivor, Viktor Frankl noted holds the power of choice. But it is a choice that is only available when we actively and consciously interrupt our habits, our automatic responses.

The psychologist and meditation teacher Tara Brach has coined an acronym, RAIN, that captures a few steps that may be useful in learning a new habit to replace the automatic, unhelpful ones we engage when feeling triggered by people or events. RAIN stands for recognize, allow/accept/acknowledge, investigate, and nurture. Specifically, she suggests that we "recognize what is happening; allow the experience to be there, just as it is; investigate with interest and care; and nurture with self-compassion."

Pema Chodron for her part suggests the four Rs, namely "recognizing the shenpa, refraining from scratching, relaxing into the underlying urge to scratch and then resolving to continue to interrupt our habitual patterns like this for the rest of our lives."

Again, we can each choose our own most effective means of achieving that pause. The key is to interrupt an automatic response long enough to disengage System 1 and engage System 2, so that we can respond in a more thoughtful manner. However, Chodron cautions that it is important to not get hooked on our ability to not get hooked, because that is also a form of attachment that could have negative consequences.

Frankl advises that creating that space—wherein we can carefully consider our options and choose to respond differently— gives us a measure of freedom. It is perhaps the same freedom we achieve from severing our attachment to things in the world, to our thoughts, to our opinions, which become ideologies over time.

It takes tremendous effort and courage to let go. It may even feel like an injustice, especially when we believe we are justified and that someone else who has triggered our response, is not— when they have done something egregious and not responding means that we are letting them get away with it.

But pausing does not mean not doing our duty, not taking the actions that may be warranted in a particular situation. It gives us the freedom to do all those things but to do so in a manner that also frees you from the feelings of indignation and disgust that may influence an ultimately unhelpful type of response.

Indeed, the point is to make better choices, to choose options that we can live with, choices that we will not regret in time because they were an overreaction, disproportionate to a given event or behaviour. We could even think of it as a form of wisdom. It is, if nothing else, the freedom to develop our own wisdom in a way that incorporates compassion for both ourselves and others.

27

Accepting What Is

Much of our trouble with being alive —health problems, accidents, and violence aside—has to do with our inability to accept things as they are. This includes the world, as we find it, and the people in the world that we interact with.

Looking up at the night sky it is hard not be astounded by the innumerable stars that stare back at us. The vastness of it, the absolute wonder of it cannot but blow the mind. How in the face of this mystery can we continue with an ordinary life, with its quotidian concerns and caprices? With its petty grudges; its hurts and seeming joys? All of it pales in comparison to the ego-dissolving magnitude of the universe. You don't matter. I don't matter. Nothing matters. It's all matter.

Does this suggest that we may as well do whatever we want? Or do nothing at all? It would seem to make no difference. In the larger scheme of things, this may well be true. But on the level of everyday life, it matters what we do. Because we are part of an ecosystem. What each of us does affects everything and everyone else in that system. We can, and do, change the course of the universe; or rather, the course of the universe changes us. It is the same difference. That is because we do not stand apart, we are a cog in a much larger wheel and whether we turn, or are turned by it, is irrelevant.

One possible consequence of this is that our behaviours, no matter how extreme instances, will ultimately revert or regress to the mean, especially over the long-term (as measured by the length of our lives). Most of us, will, by definition, be average, even if for brief, intense periods we rise above or fall below the mean. Fatalistically, then, this is destiny. It is futile to try outrunning it.

And the universe seems inevitably deterministic when we consider this idea with the notion that there is no free will. The fundamental constants of nature and the laws of physics in our universe were presumably established at the big bang, and everything that has happened, and will happen, is determined by those initial conditions.

So, what is a person to do? Is there any point in setting goals? Having dreams? Trying to change our lives?

Perhaps it is depressing to think of those dreams and hopes as the results of a dice roll long ago; to think that all our seeking and striving is simply the pre-determined and inevitable unfolding of the universe, albeit subject to randomness.

Not just depressing but also nihilistic. That line of thinking could prevent us from living fulfilling lives. We must thus act as if we have some control. We should try to change while remaining fully aware of the possibility of failure, acknowledging that the probability of success or failure may have already been determined, even if there is no way for us to predict or know the outcome in advance. Or at the very least it is unpredictable, subject to mathematical chaos.

Thus, we are better off focusing on living, on the small everyday puttering that is our lives. We may be far happier accepting that— to paraphrase the author Annie Dillard— how we spend our days is how we will live our lives. Meaning that it is wiser to emphasize the systems and processes that underlie our routines, our everyday existence, over goals and outcomes, which are subject to probabilities that we can't entirely influence. And the farther out our goals, the cloudier the forecast, the greater the number of variables that need to be considered, and the weaker our ability to predict or control the outcome.

And here, too, because we are destined to be average, and our mortality is guaranteed, accepting these things could give us a measure of freedom to live our life unencumbered by our own expectations or those of others. We can then choose to be unimpeded by lust for fame and glory or greed for money.

Perhaps there is some wisdom in the serenity prayer— to accept the things we cannot change and mustering the courage, the boldness to change the things that we can, even if we may never know the difference. That acceptance includes acknowledging that the things we can change are for the most part things about ourselves, and specifically the decisions we make moment to moment, day to day, not the plans we make for the years or months ahead and perhaps not even our goals for the week.

It may also be helpful to realize and accept our tendency to do the easy things, not the hard ones. We have a penchant to opt for what the psychologist Daniel Kahneman calls "cognitive ease." These are the tasks and decisions that don't require a lot of mental effort, which arise automatically in the brain in response to a given situation. They are automatic because the brain looks for clues about the environment in which it finds itself, clues that may offer information about what is familiar or can be assumed to be the same. And if there are enough similarities then perhaps the thing that worked before can work again— the old assumptions still hold, and no extra processing is necessary.

This way, when we react, we can do so without engaging more complex mental machinery, which would slow us down and require expending more energy. That is, we fall back on old habits— the learned, familiar routines. We do what we have done before because doing anything else is harder. Call it laziness but we could just as well characterize it as efficiency or energy minimization. It makes a lot of sense from an evolutionary perspective.

This is one reason it is so hard to fundamentally change ourselves through a simple act of will. Achieving life goals is more complex than simply employing discipline and committing ourselves to do what it takes. We also must create the environment to make change possible. In fact, it may be advisable to stop obsessing about goals, about measurement— to exit the quantified life, at least the obsessive, competitive version of it that is more about keeping up with others and keeping up appearances than about anything intrinsic or self-motivated.

Changing our environment means setting it up to remove the hurdles that can prevent us from changing whatever it is we want to change. For instance, if we want to eat more healthfully, it may mean not buying processed food in the first place. If it is not in the kitchen cabinet, we will have to expend more effort to go out and get it. And often, we will resort to eating what we do have at home rather than heading out to buy what we don't. As they say about eating in general, changing our diet is preferable to dieting. The first is a permanent change while the latter is a temporary adjustment that will be harder to sustain, even if we are able to do it successfully one or more times.

Similarly, whatever the change we hope to achieve, it helps to make it as easy as possible for ourselves. For instance, if we aim to get up earlier in the morning, we could try breaking the change down into its components. We should also consider what aspects of our environment can support one or more of those components, including through alteration (for example, moving the alarm clock away from the bedside). But through it all, we shouldn't invest too much in the goal itself. We should avoid tying success to our self-worth or identity. We need not reduce our life to specific goals and achievements, many of which are only partly within our control, if at all.

Instead, we should consider the process itself as a potential source of satisfaction and perhaps even serendipity. Doing so may lead us down a different path than we intended, which in turn may reveal unexpected delights. We could also form habits, develop rituals and imbue them with meaning. These don't need to be religious or spiritual, but they must be easy to do. These must be things that we are willing to do day-after-day or with whatever frequency we choose without a great deal of thought or specific strategy.

But amidst it all, it is important to recognize that change is difficult and while it may sometimes be necessary, it is not always so or not altogether possible. We need to remember that change entails something from both us and the world. And while we may think we mostly control the former and perhaps even some of the latter, that is unlikely to be the case.

We are subject to the winds and whimsy of fate, however defined, and we often blow the way it blows. And when appropriate, accepting that is easier than fighting it, no matter how much we may have been told— and perhaps even believe— that our future is in our own hands; that we can change the world, if only we try hard enough.

This is a lesson that is not easily learned. Often it seems as if those of us who grow up comfortably have a more difficult time coming to terms with this than those whose lives have always been subject to other people's whims— those that haven't had the power to make even simple decisions about themselves.

28

Looking at What We See

Perhaps one of the biggest tools to understand our place in the world is also one of the simplest. Knowing how much of a sense-making, storytelling machine our brain is, we often don't realize that telling a different story is even an option. Specifically, telling ourselves a different story about who we are. This is not about fabulism, about making up fairy tales. But what is a white lie when you are already living a version of the truth because of your inability to perceive reality as it is—if an objective reality even exists.

But focusing on psychology rather than metaphysics, there is value in creating the narratives we want to live. Psychologists talk about reframing or dialectical thinking as one of the most powerful tools available to train ourselves to respond differently to the same stimulus. Reframing involves interpreting a person or event through a different lens. That is, it guides us to take on another perspective. A failure becomes a lesson learned, an illness an opportunity to slow down and take stock, an accident becomes a near miss that makes us glad it wasn't worse.

Reframing is re-telling, often through a less negative interpretation of the facts. It adjusts facts that may already have been distorted based on our predilection, say, towards pessimism or optimism. The point is not to adopt a Panglossian world view but rather a realistic one that leaves room for both uncertainty and hope. When the odds are unknown, the chances of a favourable outcome ought to be, at least in our own minds, as good as that of a negative one.

Perhaps nothing highlights the link between narratives and our health better than the placebo effect. Although long studied, it is still not entirely clear why or how it works. And although it is considered part of the so-called gold standard in clinical trials, where efficacy has to be shown to be better than the results achieved by taking a placebo, what's often missed is the power of the placebo itself.

Psychologist Ted Kaptchuk, a leading researcher in this area, has suggested that the placebo effect is the biological response to a perceived act of caring— whether by a physician or anyone else. It is as if an interaction with another person triggers the body's healing mechanisms and that the efficacy of it is correlated with the intensity of the other person's care and attention.

While this is a remarkable view, there are some experimental results that back it up. For instance, even telling a patient that they are getting a placebo doesn't seem to diminish the effect— for instance when treating someone with irritable bowel syndrome. And the effect is even stronger when the physician takes their time and treats the patient with warmth and attention.

But the effect seems to extend beyond the physician-patient interaction as well. Consider that people who are told that they have had knee surgery start to walk better even when there has been no surgical intervention. And athletes' performance improves when they are told that they are breathing oxygen when they are just inhaling regular air. It seems to work for a broad array of both physical and mental ailments, from backpain and depression to migraines and post-traumatic stress.

On the flip side, people who are told that they are drinking alcohol start to behave as if they were drunk even when they have only been given some other non-active liquid.

It seems, then, that training the mind— to the extent that it can be trained— is indeed an important way to both feel and be better— to, in essence, be healthy. Of course, as with many things, rewriting our stories is easier said than accomplished. But here again, habits help. First, we can recognize when we are adopting an unhelpful perspective and then start working to reframe it. Repetition should make it progressively easier, until it becomes second nature— that is, until it becomes a habit; a System 1 response that doesn't call on the much-harder-to-corral efforts of System 2.

The thing to remember is that we are constantly telling ourselves stories about who we are. I am this or that type of person; I like this, I don't like that. I am a visual learner. I am an auditory learner. I am a no-nonsense person. I tell it like it is. I have integrity. I am honest and trustworthy. I am a good person. I am a fraud. I don't know what I am doing. I am dumb.

Our stories are often exaggerations, influenced by what we may be thinking or feeling at a particular moment, which in turn may be at least partly determined by where we find ourselves— geographically, socially, biologically— at a given point in time. The words we use— in our own heads, even if we don't always say them out loud— have the power to change how we feel and the actions we decide to take. Reframing builds on the idea that changing the words and changing the frame of reference can change what we think and feel.

This is not to say that we will suddenly acquire new powers— be able do anything, realize anything we set our minds to— as countless platitudinous commencement speeches and self-help seminars will have us believe. There are limits—things that are too far out of our control for us to exert any meaningful influence. These are both physical and psychological. Just because something has to do with our own behaviour doesn't mean that we have an infinite capacity to change it, to wish it away.

There is still much that we don't understand about the extent of our free will, the degree to which we can control our own behaviour. Regardless of whether we have any agency, there seems to be more benefit than harm in pretending that we are in charge of ourselves, that we have the ability to change. In that context, reframing is thinking about our current situation in a different way. It is a chance to depersonalize it, to try on a different way of being; consider our state of being from someone else's perspective. But to be effective, the new frame must be relevant to us personally. Ideally, it has to carry some emotional import to be effective and sustainable. That is the only way it can be compelling enough to change reality, because that's what we are actually doing— we are altering our perceived reality, the very model our brain uses to understand its world and take actions.

By casting an experience in a new light, by focusing on the opposite of whatever we may have initially and automatically thought, we are slowing down our mental processing. We are engaging System 2 to rewrite the story. Sometimes, it may even be helpful to write things down, to get the story out of our heads and on to a page, to make it physical and therefore more real. It may also be useful to start acting as if we believed the new story, to make physical motions that demonstrate the new frame— to let the body lead the brain.

Psychologist Elaine Langer and colleagues conducted a series of experiments to demonstrate the power of reframing. One of the most widely known is an experiment from the late 1970s where she put eight elderly men in a cabin in the woods for a week. The cabin was decorated to resemble a home from 1959 when the men had been in their twenties. While at the cabin, they were instructed to act and speak to each other as if they were indeed in early adulthood rather than in their dotage.

A control group also spent time in the same cabin but weren't told to behave as if they were younger than their actual age.

When tested, the experimental group that was told to act as if they were all young men showed noticeable improvements in a range of areas, from memory and posture to dexterity and sight. They supposedly even looked younger to independent observers who weren't told about the experiment. Although the sample size was small, Langer has demonstrated similar effects with other studies where people were told to reframe their environment or their lives.

In one such study, hotel chambermaids who reported not getting much exercise were told that given the type of work they did— cleaning, walking from room to room, bending over frequently— they were actually engaged in serious physical activity, equivalent to or greater than that recommended in the U.S. government's health guidelines. Compared to a control group that wasn't told to reframe the same work, the maids in the experimental group lost weight and improved on other measures of health, including body mass index and hip to waist ratio, even when controlling for other variables in a regression analysis. As with all such analysis, it is important to remember that correlation does not always indicate causation.

Regardless of the broader applicability of such ideas to everything we do, the results of these experiments do suggest that we may be able to think our way to a different state of being— to the point of realizing tangible, physical benefits. And if physical benefits are possible from reframing, why not psychological ones? If what we say to ourselves changes our perception of reality then the language, or specifically the words, we use to describe our experiences and the world around us could also change our perception and indeed the experiences themselves.

This is what psychologists such as Lera Boroditsky have suggested. She and her team have conducted experiments that seem to suggest that grammatical and syntactical structures of different languages influence how native speakers think and perceive. Her work with anthropologists has shown how time and direction are experienced differently based on what words are used to describe related concepts in different languages— or intriguingly— even the definition of truth and lie. Further, the information content of a word can vary depending on its grammar. For example, gendered nouns and multiple tenses can yield additional information about an event, such as whether a book was fully or only partially read. This makes sense if we consider that even in common, and perhaps relatively less (grammatically) complex languages like English, the definition of a word and the tone used to express it can change the way it is heard and interpreted by a listener.

For an example of how space can be experienced differently, consider the Kuuk Thaayorre and Guugu Ymithirr aboriginal societies in Australia. They and other tribes on the continent use a cardinal (that is geographic, North, South, East, West) frame to navigate their space, even indoors and in relation to themselves, rather than the egocentric ones many of us are used to (to my left or right; behind me or in front of me).

This way of seeing their world, seems to have given Guugu Ymithirr and their kin a superpower— they can quickly and easily identify the geographic direction they are facing, no matter the location. They can do so even when blindfolded and confined to a dark room without windows; and after being spun around a few times. But this does not appear to be an evolutionary adaptation unique to them, or even a special skill gained through supernatural means. Rather, from a young age they are accustomed to using geographic directions to the point where they pay closer attention to, make note of, and track environmental cues (where the sun is, what feature of the land is located where) than most of us are used to doing.

This is not unlike the suggestion that speakers of tonal languages are better at music, because they tend to have better than average "perfect pitch" based on early exposure to the meaning of a word changing based on how it is said— the word "ma" in Mandarin and Cantonese, for instance, could mean mother or horse, depending on the pitch.

All this suggests the critical importance of perspective in shaping not only our view of the world but also of ourselves. And beyond being the basis for our ability to navigate the environment and our relationships, it is also a tool for change. That is, it can help us change how we see situations, ourselves, and other people—for better or worse. The choice may be up to us.

29

The Price of Everything

Unlike the body and memory, values are perhaps in large part a social construct. Nevertheless, they define us in our interactions with others and make us part of who we are.

Values are not about morality— about what is right and wrong— but rather about choices and trade-offs. They implicitly reflect our priorities in life. In a practical sense, values represent ideas and our emotional response (positive or negative) to those ideas. Thus, a value like justice is defined both by our ideas about justice representing fair treatment of all, as well as by our belief that justice is a good thing, and that injustice is bad.

The economist, Amartya Sen, defined freedom as the ability to pursue life in accordance with our values. But perhaps the qualifier is that this may be true if we do not harm others in that pursuit. The philosopher, John Stuart Mill's definition of the same term in "On Liberty," seems more fitting. For him, liberty is "doing as we like, subject to such consequences as may follow, without impediment from our fellow creatures, as long as what we do does not harm them even though they should think our conduct foolish, perverse or wrong."

Yet, for most instances and purposes, we do not actually practice freedom. So much of our lives are spent trying to be the people that we think we should be— that other people or society expect us to be. We mask our true views because they may not be socially or professionally acceptable; or because we feel ashamed, fear reprisal or reprimand. Instead, we espouse opinions that we believe will make us fit into our culture.

There is often a discrepancy between what we say our values are and how we say we will act in a given situation and how we actually do. The implied values are different from the espoused values. Yet, here too, we will more readily excuse ourselves—rationalize our actions—yet not forgive others so easily.

Some of this is understandably necessary—to get along by going along—as we exist not in isolation but as part of a collective, where following social norms—most of the time—benefits all of us. The economist, Timur Kuran has termed it "preference falsification," which he defines as the phenomenon wherein people either remain silent and don't express views or profess to believe the exact opposite of what they really believe. Of course, it is not a case of us against the world. We do this to each other. We are, after all, part of society and in that role, we exert pressure on others' behaviour just as they do on ours'.

Pretending with others is one thing but being dishonest with ourselves about what we really believe is counter to the goal of self-knowledge. It is better to acknowledge our deepest feelings—no matter how socially unacceptable—to not judge them but understand that we are having those thoughts and to let them go if we can. If there are things about us that we want to change, we can't begin to do so until we practice brutal self-honesty and acknowledge who we are to ourselves.

Even once we have done so, we need not be confined to a particular value system but can be open to changing it as we live in, and adapt to, people around us. This is especially the case when we make choices, particularly hard ones. Hard choices entail trade-offs, deciding which among different considerations we value more; which will have a greater bearing on our decision. And there, too, different values may take precedence in different decisions.

The thing to remember about values, however, is that they are not quantitatively comparable— they can't be simply or absolutely described as being less than, greater than or equal to each other. They are personal and subjective— not something that can be objectively, scientifically defined. This is not about moral relativism— the notion that there is no absolute right and wrong, which must be defined by the situation rather than universally. Instead, in this context, values are about choosing what matters most to us given a set of circumstances— say money or time; status or peace-of-mind. This truly is relative, because if we accept that people can make their own choices in matters to do with themselves, then the values that guide those choices are necessarily unique to them.

Take for instance a decision to do our own gardening, including the perhaps unpleasant task of weeding, rather than hiring a gardener to do it for us. If we value our time, we may decide to spend the money to hire a gardener. This will probably make us happier. On the other hand, if we value hard work and the pleasure that comes from seeing the fruit of our labours, then we may value the ability to do the gardening ourselves and derive satisfaction from the results. One choice is not necessarily better or worse than the other; yet it matters to us which we choose. Our choice has meaning; our choice will define us in our own mind, and perhaps in other people's, too.

Most of the time, our choices define us in a way that we control. While a lot of the things to do with the self are not fully within our control, this is an area that is. The self is therefore an amalgam of things that are both objective and subjective— being defined for us and defined by us. Even when it comes to aspects that may be defined for us— our early experiences, the biological mechanisms of the brain and body, with training (such as that involved in mindfulness), we can learn to influence those aspects of ourselves as well.

What becomes evident is that— at least with respect to society— we can in many ways define our role, define ourselves; determine, through our choices and our behaviour, the kind of people we wish to be. While it is often tempting to make the easier choice— taking a well-paying job rather than trying to become an entrepreneur— it is by making hard choices that we begin to create reasons for ourselves; that we begin to define ourselves.

By choosing to do the hard things— as John Kennedy exhorted in a famous speech about his decision to send Americans to the moon— we learn more about ourselves, and more about whatever it is we are pursuing rather than by taking the easier path. The philosopher, Ruth Chang suggests that difficult decisions give us agency. They afford an opportunity to define our own reasons for doing something rather than using reasons provided by others.

Ultimately, social values exist by consensus. They are not properties of the universe, laws of physics that exist independently of us. Rather, they are psychological constructs that are valid if enough people share them. Consider the value of commercial goods and services, including cars, houses, tulips, or bitcoin. The market for these assets exists because we create them through our changing preferences for them— including such choices as where to live, what to drive and how much to buy.

Economic bubbles (including the Dutch tulip mania of the 17th century) and runs on banks (where everyone tries to withdraw their money from the bank at the same time) occur when we have an excess of or lose confidence in the value of things. This is also the underlying driver of the stock market where different beliefs about the prospects for a given company at a certain point in time can drive its share price up or down.

In much the same way, social and personal values are subject to change, and we should be willing to change them as our preferences and situations change. We should be prepared to become different people, who hold different values because that is what the moment demands; and that is what we owe ourselves.

30

Picking the Right Horse

The thing we do without fail, every day, indeed every conscious waking moment of our lives, is that we make choices. Some of those choices feel deliberate, supported by rational thought, information, and analysis. Others seem to be involuntary— almost reflexive responses to unexpected stimuli. But for the most part we like to believe that we are making good, sound, thoughtful decisions based on facts, not whims. At least, rational decisions are the ones we respect as a society, the ones we feel comfortable defending, even if we know we don't always make them.

The reality is that many, if not most decisions we make are not rational but are subject to a range of cognitive errors— mind bugs— that have crept into our thinking, have infected our brain circuits, as it were. They lead us to make quick, easy decisions that we deceive ourselves into believing are thought through. And with many of these decisions, we seek to make what we believe to be the best choice from among alternatives, even if we can't always fully articulate what best means. By the same token, we may criticize other people's decisions and their decision-making skills as being sub-optimal— disastrous even— based on outcomes that we perceive as bad.

Here, though, we fail to account for differences in values that people hold, which may lead them to choose one alternative versus another. We don't always realize that there are few absolutes, that there may be more than one right answer to many questions in life— especially questions that have little to do with scientific facts. We also seem to assume that our own values are static, that we would make the same decision today as we would, say, ten years from now.

Decisions in the far future or ones that involve experiences that are difficult to imagine (without having those experiences), fall into a category that the philosopher L.A. Paul calls transformative experiences. That is, experiences, that essentially transform us into different people. Even shy of such major decisions, we struggle with relatively smaller decisions, even minor day-to-day ones. Sometimes it is a fear of not making the best decision, a nagging feeling that there may be an absolute best decision that we don't have the wherewithal to figure out, or a fear that there may be better options out there, whose benefits exceed those of the options in front of us.

It seems counterintuitive to limit our choices, to accept that we may be content even if we don't always choose the best option, that there may be a whole class of "good enough" options. In the moment, we are subject to what the psychologist Daniel Kahneman calls a focusing illusion, which leads us to believe that the question we are wrestling with is the most important one— perhaps the most momentous ever— and that it is, therefore, critical to make the best decision possible.

The computer scientist Edward Fredkin laid this out memorable in what's become known as Fredkin's paradox, which states that "the more equally attractive two alternatives seem, the harder it can be to choose between them – no matter that, to the same degree, the choice can only matter less." This point can also be illustrated even more starkly by the philosophical paradox known as Burdian's ass, which dates back at least to the time of Aristotle. The paradox is presented as a barnyard tale about a donkey that finds itself halfway between hay and water, on either side. Being equally hungry and thirsty, but unable to decide which need is more pressing, the ass finds itself unable to move, and starves to death.

In some situations, we may well be that ass, paralyzed by analysis rather than by food and water. We endlessly churn through the advantages and drawbacks of multiple options. We develop complex models to evaluate what economists call utility— maximum benefit— only to find ourselves stuck for fear of making the wrong decision. It is in these situations, especially, that we need to make "hard choices." These are often choices that are not easily quantified. That is, they are not the choice between getting a little money vs. a lot of money, but more akin to the types of choices involving options that are in the same general neighbourhood as far as potential benefits are concerned. They are, as Ruth Chang puts it, neither greater nor lesser relative to each other, but rather "on a par."

In such circumstances, as when choosing between two seemingly equally attractive jobs, between potential life partners, or two different universities of equal repute, we must decide based on something else. And Chang suggests that this something is our own reasons. That is, we should consciously choose one option over another by generating the reasons that we prefer it. In effect, we are choosing who, what and how we want to be in those moments. And in that sense, choosing among options that are on a par should be exhilarating, because we are effectively defining ourselves, determining our own fates in a very deliberate and conscious way.

Ultimately, this suggests that the choice itself doesn't matter, but the act of making a choice does. It may in fact be important to our sense of control over our lives (determinism be damned); our sense of agency and well being. It may force us to acknowledge that often there is no right answer, that we can reconcile ourselves to, and in fact be equally better off with, any of the options available to us.

Indeed, some suggest that having too much choice is itself not always a good thing, that it creates stress and may in fact lead to choices that are not in our own best interest. We are so afraid of missing out on something better— whether it is choosing a romantic partner, a vacation experience, a job, or activities for our children. And it hamstrings us in ways we may not always realize. It creates a mental load that is difficult to shed and keeps us from living the kind of life that we may want to live— one free of psychic clutter. It keeps us from contentment and acceptance; from an existence characterized by freedom from the tyranny of our own expectations.

Take several examples from the work of the psychologist Sheena Iyengar. In one, she studied how life support decisions about critically ill infants are made in different parts of the world. In the United States, parents are asked to decide whether to keep their child on life support or "pull the plug." In places like France, the doctors decide when all measures have been exhausted and when ceasing life support is the most humane thing to do.

Studies of parents in both countries suggest that, overall, Americans report having a more negative experience when faced with these choices whereas French parents seem to be more at peace. And despite this, if asked, Americans don't want to give up the ability to choose. They would not willingly hand over control to their physicians.

Similarly, Iyengar's experiments demonstrate that other aspects of choice also seem to be influenced by culture. Consider that whereas Americans may have a strong preference for either Coke or Pepsi (both caramel-coloured carbonated soft drinks), or struggle to decide between them, people in Eastern Europe do not typically distinguish among a variety of soft drinks. They tend to view it all as soda and see a bigger difference between soda and water or coffee.

If we are to choose wisely, then we must first define our values—

what is important to us. We can then use such values to help us make decisions, especially ones that are on a par and can't be easily compared or quantified. And if we understand that choosing is also a way of defining ourselves, of shaping the story of ourselves, then we may be both reminded and motivated to do it more consciously.

Choosing wisely will also allow us to free ourselves of the notion of absolute right or wrong, of perfect answers to the myriad decisions we make in life. And ultimately, we could also perhaps take some comfort in the many worlds interpretation of quantum mechanics— that whatever we choose, another version (or versions) of ourselves has chosen the other option and is experiencing the consequences— good or bad— of that choice. That is, in a sense, we are always choosing all options— there is no possibility of ever missing out on anything!

31

The Limits of Reason

We would often do well to resist our inclinations when it comes to judging other people. In judging others, we take thin slices of a person, based on fleeting impressions from momentary interactions and build an entire profile in our minds. We draw on our own fears and biases to attribute intentions to them, to anticipate what they are going to do or say before they have done anything at all. We could write an entire book with the other person as a protagonist if we were to sit down and catalogue every thought we have had about someone we don't know well.

While this kind of model building is advantageous and is perhaps evolutionarily adaptive— making it easier to tell friend from foe— it is less useful in the more stable environments that we find ourselves in today. In fact, it could harm our ability to form meaningful relationships, especially with those that we initially see as being quite different from us.

Indeed, studies suggest that unfamiliar names, especially Black sounding names, are perceived more negatively on job applications. Teachers appear to have lower expectations of students with such names. On the other end, people with traditionally European names and women with male-sounding names appear to be viewed more favourably. Some combination of sociocultural and genetic inheritance conditions our thinking, which predisposes us to react in specific ways to different people and situations, without fully being aware that we are doing so. But it may be even more complicated than that.

We may in fact be influenced not just by other people in our environment but by other living organisms, most notably microorganisms within our body. One recent discovery has shown a link between the so-called gut microbiome and socialization. Biochemist and pharmacologist John Cryan and colleagues have shown that gut bacteria, or rather the absence of them, leads mice to become loners. This happened when the researchers wiped out the mice's gut microbiomes by giving them antibiotics. Looking closer, they found that without the gut bacteria, neurons in the amygdala of these mice make an unusual set of proteins that change how they connect with other cells.

When the researchers added a strain of *Lactobacillus reuteri* (*L.reuteri*) bacteria to the mice's diet, they became social and interacted with other mice again. Neurobiologist Mauro Costa-Mattioli has further shown that *L.reuteri* releases chemical compounds that are picked up by nerve endings in the intestines. The vagus nerve then sends the signals triggered by these compounds to the brain, where they affect oxytocin production. Oxytocin— sometimes called the "love" hormone— is implicated in a variety of social behaviours, including maternal response, empathy, sex, and social bonding / relationship formation. It is a key hormone in childbirth and childcare and is produced during labour and breastfeeding.

In some people, supplemental oxytocin also appears to relieve symptoms of a variety of conditions, including limited forms of autism, depression/anxiety, and gastrointestinal issues. With autism in particular, analysis of stool from children with the condition show unusual patterns of gut bacteria, which also suggests that the microbiome could, at the very least, be one of the factors that contribute to the disorder. Such unusual gut bacteria compositions have also been found in people with a range of other brain-based conditions.

At the most personal level, especially as adults, one of the most important yet often difficult part of our lives is relating to those closest to us— our parents, spouses, siblings and children. Although we are predisposed to seeing the more favourably than we would strangers, when we are amongst them, they can still feel like foes, and we may indeed treat them that way.

We expect more from people in this smallest and closest of groups and because of this we also experience more conflict with them. And with rapid changes in social structures following the industrial revolution and the second world war, our expectations of other people have also evolved. For instance, where marriage was once mostly an economic and practical affair it is now more fraught, with expectations of deeper connection, understanding and support for individual growth and ambition.

We expect unwavering support from our partners— for our egos, our identities and our aspirations while at the same time expecting them to be satisfied with the people, we believe ourselves to be. But these things are not always aligned or indeed compatible. To achieve our goals, we may find ourselves trying to become something we presently are not, as we strive and perform in ways that may historically have been incompatible with our identity.

Of course, this can, and often does, go both ways. Our partners too expect similar things from us, although this will vary widely from relationship to relationship. And navigating these contradictions— the satisfaction of practical, sexual, romantic, economic, and other needs from our partners— and the resulting conflicts, has become the work of modern coupledom. Where once most of us wouldn't have had the luxury to dream of such things, today these are simply starting points in a relationship. We expect our better halves to be our soul mate, greatest supporter, and co-parent, among other things. We want them to be our everything.

Thus, it is no wonder that so many romantic relationships fail— the likelihood of one person fulfilling all our varied needs is unrealistic and impossible. And perhaps it extends beyond spousal relationships to even what we expect from our parents and children. We expect our parents to have been perfect, to have made no mistakes and done right by us in all ways. When they inevitably fail, we blame our childhood environment, we go into therapy to repair the harms— real and imagined— that they may have done.

With children, too, we have come to the view that all the eggs need to be in one basket. We convince ourselves that our children should experience and will achieve everything possible. We want them to fulfill every dream we have or could dream of having. We very much want them to achieve the things that we ourselves may never have been capable of, but which other people have realized.

We believe that they can scale the highest peaks, if only we invest the necessary effort, time and resources. What we seem to be after in most of these cases is vicarious personal fulfillment. It is a shift of focus from things outside of us to what is going on inside our heads— to our own individual hopes and desires, realized through our offspring.

This appears, at least in part, to correlate with the world's growing affluence and related increase in the time available for leisure. We take the extra time to dwell on ourselves, on our neuroses, our shortcomings, that which we don't have, all of which leaves us wanting—wanting more. To counter this, we could instead try to fulfill our own needs by thinking about how we can meet other people's. By taking the focus away from ourselves, we may again regain the time lost to unending self-exploration and the interminable pursuit of self-improvement.

With parents, we could attempt to better understand them, especially childhood memories— however real or imagined— that shaped their own understanding of who they are and who they wanted to be when they were young. We could also look forward, try to understand our parents' hopes and dreams. This is especially useful when we are grown and start to think that our parents' lives have been fully lived, mostly spent, with not much to look forward to, beyond death. This is not to suggest that we can develop a perfect relationship with our parents, or really, anyone else; only that we can try to improve it, if that is something we desire— and it isn't too difficult to start.

Similarly, with children, it is sometimes hard to maintain perspective when they seem to be on a deliberate quest to annoy or upset us; and perhaps even to challenge or humiliate us. But it may help to understand that much of that behaviour is likely developmental and often the best thing that parents can do is to maintain their own composure— and compassion. We could just be around, rather than trying to monitor them or actively connect with them, especially if they don't yet seem ready for it.

When our children—especially teenagers—are in a talking mood, it may be useful, at least sometimes, to speak of things other than their behaviour. This could include questions like what things they would like to do when they grow up, where they would like to go, who they wish to become. We could ask them to tell us something new and interesting— whether it is about the world or about themselves. And as hard as it may be, we could try to listen rather than to talk; try to take the long view—to think of a time when our present relationship difficulties are behind us.

With our partners, our so-called "significant others," we could try to make them feel understood and valued, beyond being loved. We should, by all means, continue to disagree or fight, but do so in a constructive manner and seek to de-escalate— to do the hard thing of showing our vulnerabilities and our generosity of spirit. And we could let go of the need to have our partners be our one and only, our everything, and instead try to fulfill our needs through a wider variety of people— including friends and other family members— along with investing time in solitude; in learning to enjoying our own company.

Here, as in everything to do with relating, there are no easy answers, or perhaps no answers at all. These are areas that have been more inscrutable than the physical phenomenon we see out there in the universe. At least, the latter yield to inquiry, and we, as a species, have had some notable success in figuring out the so-called laws of nature. Yet we have largely failed to codify the laws of human behaviour.

While theories abound, we are nowhere near being able to predict human behaviour with the accuracy with which we can predict the flow of electrons in a circuit, or— as flawed as our forecasts can be— the weather. Perhaps the variables are too numerous or perhaps we are self-delusional to believe that we could ever understand ourselves— that a brain would have the capacity to understand itself in any truly objective manner.

32

Coming to an Understanding

It is evident, now, that we see neither ourselves nor other people clearly. Our view of others is based on numerous observations and assumptions, filtered through our biases. We then put them into categories for easy sorting— people we like and ones we don't; those like us and those that are different; one of us, one of them.

These are, at best, evolutionary adaptations, which are much less useful today than they once were. Today, we have the information and reasoning ability to see beyond these System 1 driven first impressions and we live in an environment with relatively fewer human-to-human threats (leaving aside, for the moment, the larger human-to-environment threats).

We can recognize that we are, by and large, capable of the same missteps and errors of judgment as other people. We know that our behaviour is influenced, to a significant extent, by our situation. As the Spanish philosopher Jose Ortega y Gasset put it, "I am, plus my circumstances."

This insight— that there for the grace of whatever god you believe in, go I— should make us more humble, less harsh in our judgment of other people's failures; or at least less judgmental of their supposedly wrong actions. It should make us more compassionate. We could thus adopt the stance of the Roman-African playwright, Publius Terentius Afer (better known as "Terence"): "Homo sum, humani nihil a me alienumputo," meaning "I am human and nothing human is alien to me."

And if we pair that with a variation of the Hippocratic oath that physicians swear to, it may offer a possible path to doing less harm. We can limit harm by erring on the side of compassion. It is, by some accounts, preferable to empathy, which entails adopting someone else's state of mind— feeling what they feel— commiserating to a point that may incapacitate and render us incapable of meaningfully helping another person because we are undone by what they've been undone by.

Compassion, which originates from Latin and Old French, means to "suffer with" someone. But today we use the word to describe an emotion that entails seeing someone else's pain and feeling motivated to do something about it. Thus, compassion, it could be argued, allows us to help in a more effective way that empathy (feeling what they are feeling) alone, since it is based on a deep understanding of what someone is feeling. Indeed, the study of so-called mirror neurons suggests that when we feel compassion, the same regions of our brain—even the same neurons—are activated as those in the people experiencing suffering. In effect, we may be feeling what they are feeling. So, compassion may actually be empathy, or more appropriately, empathy with purpose.

Compassion allows us to recognize our commonality and in particular, our common experiences, fallibility and susceptibility to and capacity for suffering. It doesn't rely on broad generalizations and categorizations of people. It doesn't require judgments of good and bad. It acknowledges that all of us could be placed into practically any category, if carefully and deliberately defined, to suit whatever agenda that someone has. It admits that we are, each, fully capable of both what we would consider "good" and "bad" behaviour.

Psychologists like Dacher Keltner argue that compassion is an evolutionary adaptation, perhaps related to being a species whose young are so vulnerable for so long. His research posits a link between activation of compassion and well being. Specifically, the vagus nerve—which is involved in everything from breathing and heart rate regulation to digestion and possibly even immune response and inflammation—appears to become more active when we are feeling compassion towards a broad group of people rather than when we identify with one group in particular.

The strong "vagal response," as he terms it, is beneficial because it helps us to better regulate our internal processes, to maintain homeostasis, which other researchers have cited as a major function of life generally, and the brain, particularly. That is, the brain's purpose is to maintain an optimal balance of internal states to ensure an organism's survival.

Showing compassion also triggers neurochemical changes, or perhaps it is the other way around. Regardless, compassion is correlated with production of oxytocin, which among its many roles, promotes social bonding, generosity, and altruism. All this to say that we are primed to care, to demonstrate compassion; we are motivated to take care of others.

Further, there is evidence that compassion— for oneself and others— can be cultivated through practice. In particular, the Tibetan tradition of *tonglen*, literally "give and take," has been shown to increase feelings of well being and altruism. This meditative practice involves breathing in the world's ills—or a particular person's suffering—and expelling or eliminating it through the out breath, and in the process, cleansing it. It can be combined with an active effort to visualize sending out tendrils of compassion from us to the world, imagining those tendrils touching other people and taking away their suffering.

As unlikely and as unscientific as it sounds, such actions do have a noticeable effect on our brains. They tend to improve our own sense of well-being, including by reducing feelings of stress and by increasing oxytocin production. Beyond this basic motivation, however, we need to know how to care. Thus, we also need to seek knowledge about the other to understand their needs— not just assume what those are. And perhaps the best way to do this, which also turns out to be one of the better ways of developing new relationships, is through exposing our vulnerabilities. Through a series of personal disclosures, starting small and, if reciprocated, expanding in scope and scale, we can make ourselves known to a stranger and they can become known to us. And this in turn, is likely to trigger an affinity for them and thereby also compassion.

However, in relating to others we can practice either what the Buddhists call "idiot" compassion or "wise" compassion. When we offer someone idiot compassion, it is like empathizing. We commiserate and validate their feelings— we don't try to help them out of their current state. The difficulty is that this may be what they want in the moment and offering anything else will trigger anger and rejection.

So, it is a fine balance, but wise compassion acknowledges that the other person is feeling what they are feeling, and instead of agreeing that they have a right to those feelings, it entails trying to help them reframe the situation, to see it from a different perspective. Wise compassion, as opposed to the idiot kind, has a better chance of actually helping the person.

Of course, all this is easy when we have a straightforward relationship with the person we are offering compassion to. It is much harder when it's someone we may not like, someone who we may believe has done us harm, shown us no care. Or someone removed from us by distance, culture, religion, politics, etc. Or even someone we feel has disrespected, offended or harmed us in some way.

In such cases, we need to first acknowledge that we may not be able to bring ourselves to show compassion. But that doesn't mean its impossible to find a basis for doing so. For instance, we can try to identify and perhaps understand the set of factors that may have led to the behaviour that we find unacceptable, or which makes us feel uncompassionate. And once we have done this, we may be more willing to feel compassion without necessarily finding the behaviour itself acceptable or appropriate. That is, we can try to separate the person from their behaviour.

We could also refuse compassion by saying that someone actively chose to act with intent to harm. But the reason they acted as they did may be much more complex and difficult to untangle. And we cannot rule out the possibility that their behaviour resulted from a particular set of experiences and genes, which may continue to keep them from changing their behaviour. They are, plus their circumstances, and neither we nor they can escape our biology or the laws of physics.

Thus, a bit of perspective gathering may help us understand their circumstances better. As the scholar and rights activist Frances Kissingly has said, "You've got to be willing to risk in order to make change. You've got to approach differences with the notion that there is good in the other" and you have to have "the courage to be vulnerable in front of those [you] passionately disagree with."

For as much as we may pride our own species for having theory of mind— the ability to think about what others may be thinking— we are wrong more often than we are right. We can never truly know another person's mind without asking them, and even then, our other unique human ability— language— can get in the way. All of which should give us pause. It should stop us from dismissing others too easily. It should prevent us from writing them off as irredeemable, as not worth our time and attention.

The antidote, if we want one, may be to listen— even when we are in no mood to listen, when every fibre in our body rebels against the injustice of listening to someone who won't listen to us. That act, once we get past our resistance to it, can be emotionally powerful. It can make us feel free because we were able to overcome our own doubts and do something for someone else—we were able to do something altruistic.

Indeed, if altruism is an evolutionary adaptation— a programmed response— then triggering it offers the full range of benefits that it was designed to produce. Our vagus nerve will be activated, slowing down our breathing and heart rate, making digestion easier, our immune system will be better regulated, helping us to keep inflammation— and therefore death— at bay. Our oxytocin levels will also go up, making us feel well, content, overcome with a sense of pleasure and feeling for our fellow humans.

Compassion, then, is a survival mechanism. It is a means to thrive as biological organisms— to be evolutionarily fit and successful. In other words, it can help us to be a success in the world! We will have arrived. We will have realized our full, human potential.

At the same time, we should acknowledge that kindness is hard. Compassion is hard. And we may never be able to practice it consistently; but practice may help.

33

Relating

The psychotherapist and author, Esther Perel, suggests that the quality of our relationships determines the quality of our lives. And whatever the scientific validity of that statement, most of us can attest that there is some truth to this.

Certainly, as social animals whose recent evolution has been driven more by culture and society rather than directly by biology, our relationships with others play a major role in our reported happiness and health. Despite this, one of the enduring ironies of our lives is that we are terrible at understanding other people. We invariably fail when we try to determine their motives, predict their behaviour or understand their overall state of mind.

At the same time, we are almost as equally bad at explaining ourselves to others. In both cases we simply assume too much, about others and about what they understand of us. Indeed, most of us suffer from what psychologists call the "transparency illusion." We believe that we are an open book, that our feelings are readily available to and interpretable by those around us. We think that they understand our desires, our moods, and our overall state of mind without these being communicated clearly and explicitly.

Perhaps some of this owes to what Daniel Kahneman and others call System 1 and its preference for cognitive ease. We don't want to do the hard work of explaining ourselves or of trying to understand others— it is easier to presume, to make unsubstantiated assumptions. This tendency is aided by another type of bias, known as the primacy effect, which leads us to use brief, early encounters with other people as the basis of our entire idea of them, even though we may have only glimpsed a small part of their personality.

Building relationships is hard. It may even be frightening. We risk everything— rejection, misunderstanding, invalidation, embarrassment. Everything in our being fights against exposure, stops us from revealing our vulnerable, soft underbellies, lest someone take advantage of us, harm us in some way. We are afraid even if the potential harm is more emotional than physical. Indeed, in some ways we may fear emotional pain more than physical discomfort.

To overcome it, we need to start by being vulnerable to ourselves, with ourselves. We need to practice "brutal honesty," by acknowledging our fears and shortcomings, admitting our failures and embarrassments. Once we have completed such a self-confession, we are more likely to be ready to start opening ourselves up to others.

Relationships are ultimately ongoing negotiations rather than point in time ones to reach a definitive agreement. The deal is never done but this doesn't mean that the parties are dissatisfied with the outcome. It is as if one long negotiation were chopped finer and finer, into a series of micro negotiations. And as we undertake these discussions, as we conclude one and initiate another, we are also creating a narrative, a story about our life together— regardless of whether the relating in question is romantic, platonic, or any of myriad other types.

It is the stories we tell that create a sense of camaraderie. They are what fortify us before we go out and face the world, and the others in the world that we don't have as close, as friendly a relationship with.

This could include our colleagues, our boss, the people who provide services, sell us things; the random strangers that stare at us a little longer than is polite, and even our perceived enemies. Stories help to define who is with us and who is against, and even who is us and who is them. Of course, most of the time, we are us and they are them—the lines are very clear.

But our story— really stories, as we write different stories with different people— can run away with us; or rather, we can run away with them, if we are not careful. When the stories focus too much on us and create a narrative of us (and perhaps a few allies) against the world, they detract from our desire to invest in other people.

When we believe that we are the engineers of our own self and success, we start to believe that we don't need other people, even if they may need us. We start to blame them for their failures rather than seeing the complexity of factors that lead to different life outcomes for different people. We don't acknowledge the role of chance in all our lives.

And that would be a shame, because being social animals, we derive much of our happiness from making other people happy— whether that is a spouse, children, friend or stranger. It is that little shot of pleasure we get from a good deed that often lasts and lasts. Research suggests that the extent and strength of social connections someone has is one of the biggest predictors of their well-being into old age. The more supportive such connections are, the better we fare, the better we get at slowing brain ageing and preserving memory and avoiding diseases such as dementia.

But beyond or before old age, supportive communities have helped people where other types of intervention have failed. Whether or not we think of groups like alcoholics anonymous (AA) and its cousins as cults, they do appear to be effective for many people that have been unable to break the cycle of addiction in other ways.

Further, for conditions such as depression where drugs and even non-drug treatments like cognitive behavioural therapy have been unsuccessful, supportive community groups like the Hearing Voices Network appear to have had much more luck. This network offers a forum for people struggling with mental health problems to talk to each other, share their stories, and to offer advice and ongoing support.

Millions of people have managed to either recover entirely or have found ways to manage their conditions to a point where they can lead mostly normal lives. This suggests that we may have underestimated the power of social interaction and inter-personal relationships; their power to heal, their status as viable alternatives to traditional medicalization, which emphasizes treatment with drugs or talk therapy.

However, accepting and acting on the importance of social connections may, in some instances, require suppressing individual needs, desires and perhaps even freedoms. And it is not a trade-off everyone is willing to make. And here, there seem to be cultural differences. Based on brain scans of people from traditional Asian and Western societies, it seems that where Westerners tend to emphasize the role and needs of the individual and their rights, East Asians tend, on average, to be more collectivist— giving group needs precedence over individual ones.

While this also plays into longstanding stereotypes about East and West, some have suggested that there may be genetic differences that account for the diverging social orientations. However, this has not been proven and more recent theories suggest that these are learned behaviours, possibly even influenced by the traditional environments of different peoples.

According to this view, those who have had to fend for themselves in hostile environments, like the American West during the frontier days of the 19th century, tended to become more individualistic because of that experience. On the other hand, those that have mostly lived in stable environments with established institutions and supports could readily expect help from their neighbours and thus tend towards collectivism. There are some intriguing examples that support this theory, including the American West, Japan's Hokaido Island and wheat growing regions of China's Yangtze River where the sometimes-harsh environments both enabled and perhaps entrenched self-reliance over cooperation.

Although it seems evident that we need other people, no matter how individualistic we are, this does not mean that we must entirely give up our own needs. We do not need to become selfless in all ways, at all times. In any case, that is practically impossible even for the most altruistic among us. Besides, it isn't useful to entirely ignore our own needs and desires to attend to someone else's.

It may help to think of it this in terms of the safety instructions routinely given on an airplane— put on one's own oxygen mask before helping someone else with theirs, even if that someone else is a child or elderly person. The rationale is that if we delay putting on a mask, we may become unconscious and completely useless to everyone else.

Further, while a social orientation appears to benefit everyone, we should be mindful of "othering" - the tendency to create in-groups and out-groups, to marginalize and ostracize. As the social psychologist Jonathan Haidt put it in one of his principles of moral psychology, "morality binds and blinds." And indeed, morality seems to be deeply tied to our identification with specific groups— "good" and "bad" people as it were. But in aligning ourselves with one group over another, we tend to think less critically, accept things from within the group on faith while subjecting anything from outside to more scrutiny— and biased assessment.

Again, as Heidt and his collaborator Greg Lukianoff put it "acknowledging that the other side's viewpoint has any merit is risky—your teammates may see you as a traitor." The antidote, according to French philosopher Bernard Henri-Levi is to "preserve, inside oneself, against all forms of social pressure, a place of intimacy and secrecy into which the greater whole cannot set foot. When this sanctuary collapses, machines, zombies and sleepwalkers are sure to follow. "

Thus, it is a fine balance between preserving our independence, our ability to think critically and act individually and recognizing our interdependence and need to work together to achieve common goals to benefit society as a whole. Perhaps one way to do this is to occasionally go against our impulses— to think and do for ourselves when we feel the need to give in to the group's needs; and to give in to what's best for the group when we feel the need to attend to ourselves first.

And it may be easier when we realize how little effort it takes to make someone feel included and recognized— to make them feel like we believe they are a fellow human, our equal. A simple note of gratitude or a kind word or gesture can go a long way towards maintaining social cohesion, towards making people feel like included— that they are part of the group, not apart from it.

34

What We Must Do

We feel compelled to act, to do something in the world. To not sit back, be lazy, consume. We are told to take charge, to assert ourselves by all manner of idioms: seize the day, take the bull by the horns, make hay while the sun shines. Time and tide, it would appear, wait for no one. Time after all is a non-renewable resource. There is, as yet no way to get back lost time.

The hours we whiled away, staring into space, letting our minds wander— these are lost hours; unproductive ones. To think of all that could have been achieved in that time. Or at least this is how the thinking goes. It allows no room for what the Italians call *il dolce far niente*, the sweet feeling of doing nothing, and more so, of having nothing to do. We could think of it as a state of active rest, that is, a deliberate choice to rest both mind and body. To be still. A contradictory state of acting in the world by not acting at all.

Not acting creates room for emptiness, offers an opportunity for calm to enter, for silence to permeate our minds, to interject into our thoughts. If we can get comfortable with doing less than we think we ought to do, we may discover that we are able to focus more on what we do do. Then we can actually savour it. We know from Einstein's discovery that time is relative to the observer and indeed it may be something imposed by the brain, a mental construct to keep signals organized. To an extent this is evident in how our perception of time varies depending on what we are doing at any particular moment. That is, our perception of how much time has passed depends on whether we are fully engaged— in a "flow" as it were— or are feeling bored and anxious.

In our rush to do, we are often driven by a fear that's earned its own acronym, FOMO or fear of missing out, coined by the author and business coach, Patrick McGinnis, to describe the anxiety that arises when we believe that something important or exciting is happening that we are not part of. It is the fear of regret that may stalk us for the rest of our lives. The "if only" and "what if" questions that we ask ourselves and that others, we think, may ask us. Perhaps it is also a fear of being left out from whatever group we desire to be a part of.

This makes us rush into things, into making decisions, without accounting for the value of those things— value to us or to others. We do not stop to consider whether the thing we are doing is consistent with what is important to us, whether it will ultimately make us happy or satisfied. For instance, we may decide to participate in something because we believe everyone else is doing it, without pausing to consider our dislike for driving (which may be required to get there), or our desire to engage in that particular activity.

If we are to be free of such compulsive doing, what is it that we can, or even must, do? Are there right things to do and if so, is there a way to do them right? And is there something we should strive for in doing those things. Is there a standard we should aim for, or even a level of excellence? Or should we just try to do our best, regardless of how it compares to any absolute standard that others may set or imagine there to be.

That is, of course, not an easy question to answer. It is something that we may each have to answer for ourselves rather than relying on others— whether it be people we know or institutions we trust or believe in— to tell us. And the reality is that the right thing to do will depend very much on our frame of mind at a particular moment in time. It will likely also be filtered through our experiences and our mental processing, which in turn is determined by our genetic inheritance and the environment.

Thus, what may seem right to one person may seem less so to another. And what may seem like the right way to do something may not seem that way to another. As so often happens in life, different people will have different perspectives on the same issue. Even supposedly reasonable people. That is how we end up in debates. How we end up on polar ends of an argument, fiercely defending our point of view, our most closely held beliefs, our core values.

Values are frequently the tenets on which we claim to base our actions, although often as not, we do things for reasons that are very much different than those we espouse; different than the reasons we profess to other people. Theory and practice diverge, sometimes starkly, to the point where we often, consciously or unconsciously, act against our own interest. The Greeks even have a word for it, *akrasia*, which means acting against one's own good judgement or, sometimes, a lack of self-control. The reasons why we do this are complex but may have something to do with what psychologists call cognitive dissonance or motivated reasoning.

That is, in an effort to maintain the coherence of some story we have told ourselves— perhaps even a story around which we have built our identity— we rationalize and make up justifications for doing the thing that we are doing, even when it is ultimately not in our own interest. The desire for narrative coherence is that strong, so powerful that we are prepared to harm ourselves to preserve it.

Disrupting it will take work, but it is the kind of action we need to practice when we act in the world, especially if we desire to be more aware, more deliberate, more conscious about what we are doing. But even if we succeed in interrupting ourselves, there is still a decision to be made about how we will be, what we will do, who we will try to be. Because everything we do, perhaps even the smallest, most seemingly insignificant actions, reflect on us, and may be interpreted— both by ourselves and by others— as a manifestation of our identity.

If only it were so simple. If only we could simply judge a person by their actions at any given moment and describe them as kind or cruel, heroic or cowardly, smart or stupid. But as we have seen, much of what we do is driven by unconscious forces, and if we live in a deterministic universe, as posited by some scientists and philosophers— if we truly have no free will— then what we do is not up to us. We act because we must. Because we cannot do otherwise.

But again, as we have discussed before, it helps to pretend that those questions are not applicable to us; that we do, indeed, have agency, the freedom to make decisions, to be the architects of our own failures and fortunes. To choose the future, as it were. Doing so may in fact help us to not live so much in the future, but be here, in the forever now, because that is when we most feel like we have control, when we know the most about our environment. It is the point in time when we are always making decisions. At other times we are either in the process of choosing or have already chosen.

Nevertheless, to be human and alive means thinking about, anticipating, and planning for what's ahead, even if only from a purely practical perspective. But we should make those plans with the full knowledge that they may not come to fruition, or at least not be realized in the exact ways we imagine that they will manifest. If we set our expectations based on probabilities, then we can perhaps be more confident in choosing a course of action, even in the absence of knowledge about its rightness or wrongness. It is simply what we thought was best at a given point in time.

And sometimes that is sufficient. If we are wrong, then we can choose differently— if not for the same decision, at least for a future one. However, sometimes, the wrong choice is a fatal or even tragic one and we cannot avoid regret. All we can do is accept that we made a mistake and hope to minimize or redress any harm to the best of our ability.

Regardless, we may need to acknowledge that we cannot always optimize what we do, that despite our best intentions, our most valiant efforts, the outcome will be less than perfect— possibly even entirely disastrous. Yet we must act, we must make decisions every day. Here, we can consider the concept of dharma— duty— and perhaps combine it with a variation of the Hippocratic oath to say something along the lines of: act but do so to minimize harm whenever possible.

35

Making Meaning

Meaning is perhaps not so much found as created. We often assume that there are some universal absolutes towards which we should all be striving. Whether it is standards of morality, purity, beauty, ability, intellect, or success. And we tend to think that achieving such standards will confer a measure of meaning on our lives, that it will make us feel fulfilled.

Like in Rudyard Kipling's stories for children about why an animal has a tail or why it does a certain thing, we, too, tell ourselves "just so" stories. And we do it to assume control over our narratives, to keep alive the story of ourselves—to feed it, to reinforce the foundations of our ever so fragile sense of self. We make up reasons for the things we have done, for how we have treated other people, for the life choices we have made. We avoid the cognitive dissonance associated with any suggestion that our decision may have been sub-optimal or even wrong, that we are somehow not who we thought we were.

In effect, what we are doing here is making meaning, but we don't always do so in a way that is beneficial to us, or to those around us. In Fyodor Dostoyevsky's Idiot, the character Ippolit Terentyev, reveals this insight into the human condition: "It's life that matters, nothing but life—the process of discovering, the everlasting and perpetual process, not the discovery itself, at all." This was so memorable, in fact, that years later, the English writer Virginia Woolf would appropriate it for one of her own characters.

It is this tension between a pre-determined destiny and the process of discovery, of the potential for the unexpected experience that Dostoyevsky seems to be interested in. And it is something we can relate to. Meaning suggests destiny, it implicitly posits something that is out there to be discovered, realized, conquered. On the other hand, lack of meaning is a kind of freedom but also frightening, it suggests that there is no plan, that we may be alone in life, fumbling our way through the dark. Blind people, in a dark room, looking for a black cat that isn't there, as they say.

We often shrink in the face of uncertainty. Our minds can't handle the absence of structure and purpose, so we fall for whatever charlatans have on offer. And often what they are selling is a mirage of security and predictability, the certainty of a perfect after life, for instance, of rewards in heaven, promises of being saved if only we submit now, in this life. If only we accept some unproven truth about the universe.

And it is a tempting offer, for sure. Who wouldn't trade in a meandering, confounding existence for one where the results are guaranteed, by a divine power, no less? Who wouldn't want to be saved? It is Pascal's old wager— there is no harm in submitting to the illusion, you could argue, regardless of the truth of the matter.

Why do the hard work to make meaning out of meaningless? Why even try? Better to buy the fiction, to live the lie, to find a semblance of happiness, no matter how illusory or short-lived. And there is always a new solution around the corner, a new answer to the eternal question of "life, the universe and everything."

The problem is, we cannot even agree on what it means to live a good life, a life of purpose and meaning. When we talk about finding meaning, are we speaking of happiness? Of freedom? Does one automatically imply or depend on the other? And do we have some fundamental right to such a life? If so, how? By what means do we derive such a right?

It is easy to conflate the question of how we should live with what we want. The pursuit of wants may not ultimately yield happiness or even constitute a life well-lived, even if it is satisfactory to the person living it. But does that mean that how we live should always consider other people's happiness and the effect of our actions on their lives.

It is hard to see why it shouldn't. We are social animals, after all. We depend on each other for survival, so it would make sense to consider others when making decisions that may, if viewed too narrowly, seem to impact only us. And if we look back on history, happiness was viewed more broadly. It was seen as being derived from living a moral or virtuous life. Individual pleasure or satisfaction of any kind had very little to do with it.

However, with the European enlightenment in the 17th century, the definition started to shift to a focus on individual measures. And more recently, with the widespread influence of Western culture on the rest of the world, happiness has become associated with more narcissistic pursuits, with a degree of solipsism. We are, now, much more concerned about individual rights and liberties. We have expectations of other people based on who we think we are; we feel as if the world owes us certain things, especially if we have earned them. These are things that aren't owed to those that haven't made the effort that we have. It is all quid pro quo when it comes to happiness and meaning.

We pursue life, really the quantified life, as if it were a contest, where there are always winners and losers. A zero-sum game, if there ever was one. And our objective, naturally, is to be in the class of winners—to possibly even become masters of the universe, if we are lucky and do it right. To this end, we pursue individual results rather than collective ones. We focus on our own optimal productivity, the perfect body, the best outward indicators of success.

What if, instead of looking for validation from the world, we were to find a comfortable point of mediocrity? What if we simply focused on minimizing harm and perhaps, beyond that, on trying to be helpful where we can and derive purpose and meaning from the quotidian. What if we were to manufacture meaning from moment to moment rather than finding some ultimate and all-encompassing version of it?

Such an approach, one might assume, is guaranteed to produce mediocrity. And no one wants to be described as mediocre. No one wants to be just average. We each seek to distinguish ourselves, to differentiate, to stand out. We want to be exceptional. We want to be better than everyone else, because that sense of superiority is grist for the ego. It is a drug that just barely regulates the insatiable appetite for validation.

Accepting our life as ordinary, average, and mostly insignificant feels like an admission of failure, especially in an environment where the messages are consistently about being the best, about achievement, about passions and dreams and the relentless pursuit of things we all desire. It partly feels that way because other people have those things and partly because we believe we deserve them, that we shouldn't be denied a piece of the pie.

But what is that pie exactly? Is it even a pie, or something worth having? And having gotten a piece, who and where will we be? Superficially, we say we value failure, that we learn from it. Yet we do all we can to avoid it. This is often because we are invariably punished for failure, sometimes even condemned—especially if it is seen as somehow moral in nature, not a match for the attitudes of the moment. We are afraid of being less than perfect.

A lack of ambition is seen as capitulation, the provenance of people incapable of scaling the heights, of achieving greatness. They are deemed sub-standard company. It is as if we have settled for less if we don't all aim to be triathletes and creators and entrepreneurs and CEOs. We are seen to have come up short if we don't try to be the best at cooking, thinking, looking, traveling and a host of other things.

But is mastery everything? Is it even possible for everyone in every endeavour? Or is a recipe for disappointment, recriminations, and despondence? Too often, we set standards based on what we perceive as other people's successes, without seeing all sides of them. Very few people can claim to be multi-talented and successful in everything they have tried; and satisfied with all of it— no matter the hagiographies. It is unrealistic, impractical and, on the whole, unnecessary.

Why should we attempt mastery to such an extent, or at all? Why is an average life not enough for us? Why does success and perceived impact on the world equate to meaning but completing a biological life cycle, from birth to death, does not?

Perhaps it has to do with that thing which we believe to be our superpower as a species, even if we aren't the ones who do it best—being social. But it is not just sociality, it's culture that may be the defining feature of humanity— the informal set of practices, artifacts and beliefs that codify how we interact and how we store and pass on knowledge.

Biological success— that is, reproductive success— has always entailed competition, primarily for resources like food and water and for mates, although it has also included competition for the things that give us access to resources, like say, fertile land and the tools needed to develop them.

But in the modern age, we have moved beyond competition for such basic resources, to competition in a cultural context, which has much less to do with survival but perhaps still something to do with passing things on to the next generation. Except now what we seek to pass on is not just genes, but maybe more importantly wealth. This makes sense if we consider that the best way to secure resources for ourselves and our kin is by having the capacity to own or buy them.

So, is the pursuit of sociocultural success— be it fame, fortune, acclaim— ultimately a search for the means to secure wealth? Perhaps so, but the question is whether success, so defined, is necessary for mere survival or even the ability to pass on our genes.

It would seem not. Think for instance of Neolithic times, when we were engaged in the most basic form of competition—for food, water, shelter, mates and perhaps fire. If the only way to ensure these was to reach the pinnacle, then that may have entailed eliminating as much of the competition— that is, other people— as possible.

However, this clearly did not happen, societies found some semblance of a balance between competition and cooperation, which may have been different from place to place. And we can readily see that this continues to the present day. There are times where we will go to great lengths to secure resources— even resorting to levels of violence unimaginable in Neolithic times— but most of the time we settle on much less intense competition that includes a significant degree of cooperation. Indeed, in The Dawn of Everything: A New History of Humanity," the archeologist David Wengrow and the anthropologist David Graeber argue that equality, equity and cooperation have long been features of human sociality.

Therefore, it is not necessary to be the standout, the best; to be perfect and achieve perfection in everything we do. We do not need to dominate, even in our own little social circles, if not in the world as a whole. We mostly need to find ways to co-operate and compete just enough— and that level of competition does not entail a lot of effort in these days of governments and laws and public institutions. We have collectively decided to contribute (through taxes) to take care of each other, though the degree to which we do so does vary widely from nation to nation.

In some places, we do the bare minimum, believing that individuals need to be responsible for themselves and should have the freedom to use their resources without being subject to the tyranny of the state. In other places— and these are not all socialist or communist societies— we have decided that we will indeed give the state a bigger role in distributing wealth, even if it means taking some of that wealth away from people who may have earned it purely through their own efforts (though perhaps it is something of a myth that anyone at all is entirely self-made, that they did not benefit from public investments and other people's labour).

Regardless of where we live and short of becoming entirely dependent on the state, we can do just enough to fulfill our duties to society and to our family, without having to reach the pinnacle of anything— professionally or personally. If it is not necessary, and if we do not have a compulsion to do so, then why pursue so-called excellence? Besides, who defines it? And should we subscribe to those definitions? Live by those standards? Strive to achieve something because it seems vaguely important or because other people are doing it?

In other words, we should examine our motives. And that may not be as easy as it sounds because it is hard to separate what conclusions we have reached through our own deliberations and what we have accepted as received wisdom. Chances are, if we stopped to think about the many things (including the way we define success and failure and even our values) we believe are derived and assumed unquestioningly. We do not always interrogate them for ourselves to decide whether it makes sense to subscribe to the same ideas as those around us.

Without self-inquiry, we are bound to misunderstand our motives. We will be inclined to do things that may ultimately neither benefit nor fulfill us in any substantive way. Instead, we may just be going through the motions and perhaps even expending great energies in futile pursuit of ideals that are not our own—pursuing ideals that don't need to be our own. We do this because we are often not comfortable with stillness. With silence. With the pursuit of nothing. We do not believe that simply being in the world is enough, that dharma is a sufficient reason for doing things. We cannot easily accept that there may be no grand purpose, no end that needs to be achieved.

It seems risible and implausible that existence may be an end in itself. This is especially so because we are human. A straightforward lifecycle may be sufficient for a bacterium or even a lion, say, but not for a human. This is the argument from exceptionalism, which easily gets extended from the species to the group to the individual. It happens to the point that each of us feels as if we ought to be exceptional. That even if we are not now, we should be able to become unique, be able to place ourselves in a category of one. What we ignore are the side effects of such mindless acceptance. We fail to see the downsides of such ill-advised pursuit of illusory perfection, of chasing after perceived ideals and idols. It can, for the many, only lead to tragedy, even if a select few— by dint of mere chance— achieve what they seek and then preach the "if I can do it, you can, too" platitude.

We are unable to live in the moment, to be satisfied with the now, without regard to what may be coming next. We have an insatiable desire to predict the future, as bad as we may be at it, and to plan for it. This makes some evolutionary sense, if we consider that a degree of planning is useful in deciding where to search for food or seek shelter; how far to travel away from home so we can return before dark.

If plan, we must, and if planning is unsatisfactory in and of itself— that is, planning for the sake of planning— then we may need to find a way to marry planning and meaning. We could call this purpose— the notion that we are planning to achieve something that is meaningful to us at the individual level. This, despite the fact that that life itself may have no real objective meaning. This, despite the possibility that life may be the straightforward result of a non-teleological process that emerged from some basic physical laws. This, despite the possibility that life begins and ceases without any consideration for the value of its existence, or regard for its continuation or purpose.

Such philosophical perspective often induces a sense of despair. It may lead to despondence and self-annihilation because we simply cannot imagine living without meaning or purpose. We need to matter to continue living. Or at the very least we need to feel as if we matter, regardless of whether we do or not.

Given this state of affairs, the philosopher Bernard Williams has suggested that each of us needs what he calls a "ground project"— our reasons for being alive; our purpose, if you will. These ground projects are inextricably intertwined with our identity, with our relationships and passions, with the cornucopia of things that makes us who we think we are.

Although Williams' definition was in part a response to what he perceived as the tyranny of utilitarianism, which seeks to maximize happiness without regard to subjective or personal considerations, it can be seen as going beyond that. By speaking to a very subjective view of the self and what it means to be true to that self, he offers us a way to find meaning in a possibly meaningless universe.

It is important to distinguish ground projects from traditional notions of success and achievement. We could have a ground project that does not seek to make a lasting and noticeable impact on the world; or one that entails fame or fortune. Indeed, ground projects are "deep commitments" that we make, which may guide many aspects of our lives. In other words, they encapsulate our values, the set of rules that we have decided to live by— ideally ones that are not just externally imposed but internally derived through self-interrogation. We consciously choose our ground project, which becomes the reason that we pursue it. It is the thing that justifies itself.

In the final analysis, we are not born but made. Moment by moment, by a brain that is using an imperfect but sophistical model of its environment to keep us alive. And we are made by us, through the stories we tell, through the meaning we make.

36

A Life That's Good

At some point in our lives, many of us try to strip away the artifice of our existence, the insincere trappings of our work and social lives, and attempt to find ourselves again. We believe that there is something to be found. We believe that there is a core— a heart— to everything, something immutable, true, pure. An essential essence that makes it what it is. That, we say, is what matters. That's what is important. The rest we could do without—if we had to.

And, perhaps for as long as we have had the luxury to do so, we have been searching for that essence. We have been looking for answers to the question of who we are, why we are here and what really matters? From the Buddha to Plato and beyond, those answers have varied over time and from person to person. Some have found their answers in religion, spirituality, art, substances, or social media. Others are still searching, desperately seeking. They are nowhere near an answer.

All the same, there is one direction that seems appealing, that many follow in an attempt to boil their existence down to the bone. This is the path of science, of rationality. And at first blush, it would seem like a quick exercise. Follow the Stoics and all will be well. After all, as humans, it is survival that is essential because all else follows from the premise that we are a going concern, as the accountants would say. Everything presupposes that we will be here tomorrow; that we will be here to have and take care of our young, to continue the species, until the species runs out of road. Absent life, nothing matters.

Even love, as much as we wish for it, treasure it, value it— it doesn't matter. In the final analysis, even love is not essential. It is merely a description of a transient state of being, a momentary intercession of euphoria in a life that will be mostly dominated by other feelings. For all the profession of love throughout history, in proposals, wedding vows and between lovers and friends; for every promise by a parent to their child, about loving them forever—love is conditional, fleeting, too often complicated by feelings.

We walk around thinking that we know something about these things. That we are experts on, if nothing else, our own selves. We think that we can authoritatively opine on what is meaningful and meaningless to us; what is essential and what is discretionary in our lives. But we only think that we are masters of our own minds.

It is the body, the body alone that is essential. The body is what must live. And it is the body— a motley collection of cells—of which the brain is a part, ensures its essentialism by using its brain to keep itself alive. And even that is only necessary for a while. The body need stay alive just long enough to reproduce and raise its issues to maturity. Then it becomes non-essential, perhaps even a burden to its world, consuming resources that may be better allocated to the young.

We are loath to see ourselves this way. We deny that we are, at the core, animals living natural, animal lives. That is not enough for us. Indeed, such an admission would be failure. It would entail admitting that our essence is not that different, not special— either as a species or as individuals.

Not only do we want our identity to be essential— we want others to acknowledge its essentialism, too. We want them to say that we are different; that we are unique; that we are valued and worthy. Of course, we want this acknowledgment, but we also want to have the things that we believe are essential to our existence. These may include, among others, the free exercise of our rights, and the right to have things, to do things, to experience things.

We develop whole philosophies to describe, rationalize and defend these beliefs. Sometimes, we call them religion, sometimes we call them politics; other times we call them rights. Human rights, that is. Which, as we all know are special, just like us. These are non-negotiable, universal absolutes, except when it suits us to think and do otherwise.

Is it, for instance, essential to limit harm? Most of us would say so; yet we do harm everyday. But we don't stop at harm. We think there are many other rights worth having— we have codified them in various forms— constitutions, charters, declarations. Why only human rights? Why not rights for animals? Plants? Rocks? The Earth? Why are other living things and non-living, non-beings less essential? Why do we draw the line here but not there?

But back to us. We are the ones asking the questions. So, while we are at it, is our happiness essential?

It would seem not, if survival is our focus, our sole purpose. Food, water, shelter— these are essential for staying alive. Everything else is debatable. Yet, we do not so much debate these other things as pursue them with an intensity that assumes their essentialism. We seek meaning, purpose, mastery, fulfillment. It is what really matters, we are told. It matters for happiness, the pursuit of which, all good people agree, is essential to a good life—eudaimonia, the good life or full life as per Socrates, Plato and later Aristotle.

It is also the examined life— ask why; interrogate your values; develop your character, virtues, intellect, friends; pursue opportunities. These are the things that are supposedly essential to the pursuit of happiness. Else you have lived for nothing. You may as well have not lived at all. Perhaps that's an extreme interpretation, but crucially the Greek philosophers saw the essentials as being circumscribed by that which contributes to happiness. And they helpfully noted the specifics.

However, if we equate essential with minimal then the Greek definition hardly fits the bill. Their philosophy requires a great many things and these need to be actively cultivated— goals need to be pursued; ideas need to be debated; friends need to be engaged. Frankly, it sounds more maximalist than minimalist. Living the complete, fulfilled life is about anything but the essentials.

Millennia later, Abraham Maslow put it more simply. For him, self actualization— the realization of one's full potential and therefore "complete life" is only possible after the essentials are in place. The essentials being satisfaction of physiological needs for survival (food, water, and sundries) but also the psychological needs for safety, love and esteem. Even though he considered these elements to be at the bottom of the hierarchy of human needs, Maslow believed that we need more to thrive. For him the essentials incorporate the whole pyramid, not just its base.

Is this so? Do we really need the things that lie near the apex of Maslow's pyramid? Do we need self-esteem? Do we need love? Do we need to self-actualize? Or are these desirable but not absolutely essential to survival? For what proportion of our history have we even been able to entertain the idea of contemplating anything beyond basic survival?

A lot of things seem essential because they are caught up in ego— in our reluctance to embrace mediocrity and simplicity. With that reluctance comes the burden of living up to our own expectations, which in turn bring their companions—anxiety, dissatisfaction, ennui. The essentials, one would think, ought not to be burdensome. They must be like elementary particles, like the fundamental forces of nature. Indivisible. Stripped down to the marrow of existence itself.

The 19th century American essayist and naturalist, Henry David Thoreau described one way of achieving it in practice. His experiment in the woods was designed "to front only the essential facts of life and see if [he] could not learn what they had to teach, and not, when [he] came to die, discover that [he had] not lived." For Thoreau, this is what it meant to live deliberately and there is a certain appeal to that.

We spend our lives striving and rarely stop to wonder what all that striving is for. Could it be striving for the sake of striving? Could it be that we have inadvertently stepped on to the hedonic treadmill and can't figure out how to get off? Perhaps we have drunk too much of the Kool-Aid, glommed too much onto the American idea of life, liberty and the pursuit of happiness. Could it be that it is this pursuit itself that keeps us on the damned treadmill?

Success is in many ways a gateway drug; it can become an insatiable addiction that needs to be fed with more powerful drugs— bigger and bigger "achievements." This is what the psychologists Philip Brickman and Donald Campbell called the "hedonic treadmill." Once we get on, we can't get off— at least not easily and not without falling far. When we are on the treadmill, things that made us happy once are no longer enough. We need something more. And so, we pursue it and for a while that seems like just the thing we needed. We feel happier. For a while.

Then we convince ourselves that we need something bigger, that getting to the next peak will finally make us happy— permanently. But soon we discover that there is no promised land, only peak after peak, and in between are valleys, where we spend most of our time, scheming and striving to get to the top again. Every now and then we may manage to scale one peak only to realize that there are more. There's always something more to want, something that seems within reach. If only . .

Yet, a lot of times we do not see it as such. An achievement, a win— recognition, fame, fortune— appear to make us happier. But then that elation subsides, and we return to whatever level of feeling we had before. We do not stay at the new high. So, did we just run faster to end up in the same place?

Would we be better off— could we step off? if we could stop climbing? — if we let go of the eudaimonia and sought to meet only our fundamental needs for survival. And beyond that, if we did nothing more than care as best we could for those around us and attempted to minimize harm— is that not a good enough life? Perhaps it is far from perfect, but is it good enough?

Should we allow for the remote possibility that in accepting that our purpose may be to do nothing more than live, breed and die, we may finally find a kind of freedom that is more valuable, and ultimately more fulfilling, than any kind of fame or fortune.

Focusing on the essentials of life may finally give us freedom from, rather than freedom to, or freedom of. That is, freedom from hunger, thirst, and harm. Everything else follows.

www.ingramcontent.com/pod-product-compliance
Lightning Source LLC
Chambersburg PA
CBHW070542010526
44118CB00012B/1189